D0824776

PRINCIPLES OF MATHEMATICAL MODELING

This is a volume in

COMPUTER SCIENCE AND APPLIED MATHEMATICS

A Series of Monographs and Textbooks
Editor: Werner Rheinboldt
A complete list of titles in this series appears
at the end of this volume.

PRINCIPLES OF MATHEMATICAL MODELING

CLIVE L. DYM

Professor and Department Head
Department of Civil Engineering
University of Massachusetts, Amherst

Associate Professor
Department of Physics
Smith College

New York San Francisco London
A Subsidiary of Harcourt Brace Jovanovich, Publishers

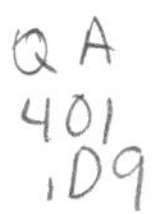

Quotations used from the following sources are acknowledged:

[Ch. 1, p. 1] S. I. Hayakawa, *Language in Thought and Action,* 1949, p. 31. Reprinted by permission of Harcourt Brace Jovanovich, Inc.

[Ch. 2, p. 11] H. L. Langhaar, *Dimensional Analysis and Theory of Models,* p. 1. Copyright © 1951. Reprinted by permission of John Wiley & Sons, Inc.

Academic Press, Inc.
111 Fifth Avenue, New York, New York 10003

United Kingdom Edition published by
Academic Press, Inc. (London) Ltd.
24/28 Oval Road, London NW1 7DX

ISBN: 0-12-226550-5
Library of Congress Catalog Card Number: 79-65441

Printed in the United States of America

To the next generation—
Jordana and Miriam Dym
William Spencer and John Allen Ivey

CONTENTS

PREFACE AND ACKNOWLEDGMENTS xi

1
GENERAL NOTIONS; AGENDA 1

PART A FOUNDATIONS

2
DIMENSIONAL ANALYSIS 11
Dimensions and Units 11
Dimensional Analysis—Motivation 14
Dimensional Analysis—The Process 17
Units 24
Summary 26
Problems 26

3
SCALING 28
Size and Shape 28
Size and Function 33
Size and Limits 38
Consequences of Choosing a Scale 41
Summary 48
Problems 49

4
APPROXIMATION AND REASONABLENESS OF ANSWERS 51
Taylor Series 51
Binomial Expansion 54
Trigonometric Series 57
Algebraic Approximations and Significant Figures 59
Validating the Model—Errors 63
Averaging—Mean, Median, Mode 66
Curve Fitting: Least Squares Method; The Continuum Hypothesis 69
Validating the Model—Testing 77
Summary 78
Problems 79

PART B APPLICATIONS

5
FREE OSCILLATIONS OF A PENDULUM 85
Results of an Experiment 85
Some Dimensional Analysis 87
Equations of Motion 89
More Dimensional Analysis 92
Linear Model of the Pendulum 95
Energy Considerations—Conservation 97
Energy Considerations—Dissipation 100
Nonlinear Model of the Pendulum 103
Summary 107
Problems 107
Appendix: Transformation of Equations of Motion 110

6
FORCED MOTION OF LINEAR OSCILLATORS 112
The Spring-Mass Oscillator 112
Building Vibration 116
Automobile Suspension System 119
Acoustic Resonator 122
Particle Moving in a Magnetic Field 125
Electrical Circuits and the Electrical-Mechanical Analogy 127

Forced Motion of a Linear Oscillator:
Resonance and Impedance 130
Summary 135
Problems 136

7
TRAFFIC FLOW MODELS 139
Macroscopic Traffic Flow Theory—
I. Continuum Hypothesis 140
Macroscopic Traffic Flow Theory—
II. The Fundamental Diagram 142
Linear Car-Following Models 147
Stability Analysis for Linear Car-Following Models 152
Nonlinear Car-Following Models 154
Summary 156
Problems 156

8
EXPONENTIAL MODELS 159
Exponential Behavior 159
Exponential Functions 164
Radioactive Decay 167
Discharge of a Capacitor in an *RC* Circuit 169
Charging a Capacitor in an *RC* Circuit 171
Inflation 172
Compound Interest 173
Growth in Demand for Highways 175
Population Growth 177
The Lotka–Volterra Equations 179
Lanchester's Law 184
Summary 188
Problems 188

9
OPERATIONS RESEARCH: LINEAR PROGRAMMING 190
Optimization via the Calculus 193
Optimization via Linear Programming 198
The Transportation Problem 204

Network Analysis 208
Summary 213
Problems 214

10
DIFFRACTION AND SCALE 217
Diffraction Geometry 218
Diffraction Gratings 226
X-Ray and Atomic Particle Diffraction 230
Sound Wave Diffraction 234
Summary 237
Problems 238

APPENDIX I
ELEMENTARY TRANSCENDENTAL FUNCTIONS 241
APPENDIX II
THE DIFFERENTIAL EQUATION $dN/dt = \lambda N$ 245
APPENDIX III
THE DIFFERENTIAL EQUATION $m\ d^2x/dt^2 + kx = F(t)$ 247
BIBLIOGRAPHY AND REFERENCES 251
INDEX 255

PREFACE AND ACKNOWLEDGMENTS

Science and engineering depend heavily on the concepts of modeling. It would therefore be useful for students of engineering and science to be exposed to these concepts early on in their studies. With this in mind, we have written this primer on the principles of mathematical modeling.

The book is organized into two major parts: foundations and applications. In the first part we lay out many of the mathematical ideas of interest to the model builder, that is, dimensional analysis, scaling, and elementary ideas of approximation of functions and curves. In the second part we develop a series of models and discuss their origin, their validity, and their meaning. These models include linear and nonlinear oscillations of a pendulum, a host of simple (linear) oscillators, traffic flow models, exponential models, operations research models, and optical and acoustical waves.

Several important features ought to be borne in mind. The first is that the two parts of the book are loosely connected: In developing the models we have not slavishly followed the outline of the first portion of the book. We chose not to do so in order to make the second portion of the book immediately accessible to the reader, with appropriate reference to the foundations portion of the book. The following matrix indicates how the chapters relate to each other:

Matrix Representation of Relationships between Tools and Models

Tools \ Models	Pendulum (Chap. 5)	Oscillators (Chap. 6)	Traffic flow (Chap. 7)	Population, economics (Chap. 8)	Operations research (Chap. 9)	Diffraction (Chap. 10)
Dimensional analysis (Chap. 2)	●					
Scaling (Chap. 3)	●		●	●	●	●
Approximations (Chap. 4)	●	●	●	●		
Elementary functions (App. I)	●	●		●		
Equation $dN/dt = \lambda N$ (App. II)	●			●		
Equation $md^2x/dt^2 + kx = F(t)$ (App. III)	●	●		●		●

A further note is that the problem sections are an integral part of the book. There are problems at the end of each of the chapters (except Chapter 1). While there are many that provide drill in certain skills, others extend the substantive information of the relevant chapter either by developing new models or by providing numerical data for a model already discussed. For example, in Chapter 3 we display in two problems the role of dimensionless groups in analyzing experimental results, in Chapter 5 we show how dimensional analysis interacts with other approaches in deriving equations for the pendulum, and in Chapter 6 we have inserted data on resonance and impedance in a variety of oscillators into the problem set.

We anticipate several ways in which this book can be used. The material in the book has served as the basis for courses in modeling taught to first-year engineering students at Carnegie-Mellon University and the University of Massachusetts. The book could also serve as the textbook for a first course in applied mathematics, such a course being specifically for majors in mathematics. Also, much of the material can be used to supplement basic physics courses in both science and engineering curricula. In recognition of these potential uses, we have tried to make the book accessible to as wide an audience as possible. Thus, the mathematics required of the reader is just a knowledge of algebra and trigonometric functions, and the ability to differentiate elementary functions. A basic knowledge of physics, largely elementary mechanics, is also assumed.

We wish to acknowledge several colleagues who read (and endorsed with varying degrees of enthusiasm) portions of the manuscript. These valued critics—at the University of Massachusetts unless otherwise indicated—include Professors Richard J. Giglio, Jess Josephs (Smith College), James W. Male, Hugh J. Miser, A. B. Perlman (Tufts University), G. A. Russell, and Paul W. Shuldiner.

We also acknowledge with pleasure the very useful reviews of Professors David Powers of Clarkson Institute of Technology and Ted Brooks of Baylor University. These reviews were solicited by our editor, Stephen Guty of Academic Press, who was an encouraging and stimulating collaborator during the completion of the book.

In addition, directly (CLD) and indirectly (ESI), we are grateful to Ms. Irene Clarke who was the barricade to CLD's office when protection was required in order that "creativity be nurtured." And, finally we owe an enormous (and probably unrepayable) debt of gratitude to Ms. Michaline (Shelley) K. Ilnicky. Shelley typed the manuscript with a combination of skills not heretofore seen by either of us, including speed, accuracy, regard for appearance, and unflagging cheerfulness and civility! תושלב"ע

Amherst, Massachusetts
September 1979

Clive L. Dym
Elizabeth S. Ivey

PRINCIPLES OF MATHEMATICAL MODELING

1

GENERAL NOTIONS; AGENDA

Models are representations of objects, processes, or anything else we wish to describe or whose patterns of behavior we wish to analyze. This book is concerned with *mathematical models*: how models are generated, how they are validated, how they are used, and how and when their use is limited. It sounds like a very extravagant agenda and yet, in analytical terms, what we propose to discuss is not all that complicated. The basic idea behind this book is in fact very simple. That idea stems from our experience that many people confuse models, and particularly mathematical models, with the physics or the reality of the problem with which they are dealing. That is, when doing mathematics in connection with a problem, it is often very difficult to remember that we are dealing with an *abstraction* of the problem that has only an indirect connection with the problem. This notion has been very nicely stated elsewhere:

> *The symbol is NOT the thing symbolized; the word is NOT the thing; the map is NOT the territory it stands for.*
>
> S. I. Hayakawa

The preceding quotation derives from a book on semantics, not from a book on mathematical modeling!

We will develop an approach or a perspective that we find useful in engineering analysis and design and in scientific investigation. Mathematical models by their very nature have built-in assumptions and approximations, and therefore have restricted ranges of validity. These ranges of validity may often be quite broad, and if the models are carefully and reasonably constructed, they will certainly be broad enough to cover the physical problems they are meant to identify. However, we often forget that we are dealing with a model, and thus

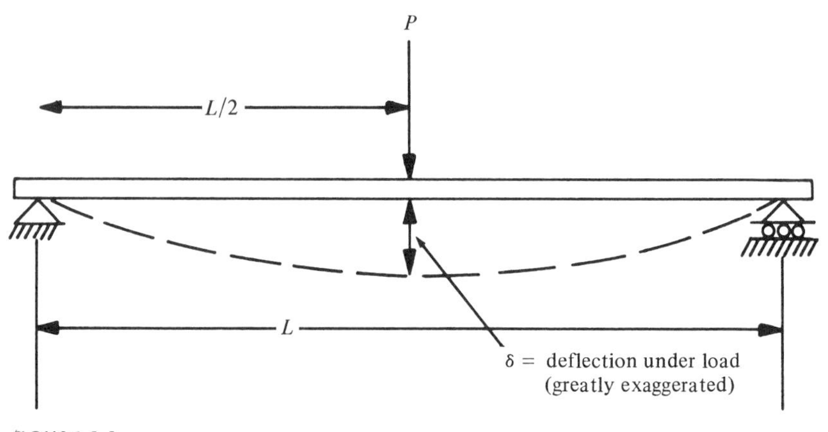

FIGURE 1.1
Deflection of a simple beam under load

we inadvertently take the model well beyond the appropriate range of validity.

For a simple example, we turn to the theory of beams. A beam is perhaps the most frequently used structural element in construction; a familiar example might be a footbridge. A simple beam is shown in Figure 1.1. The beam carries a load P at its center, and it deflects an amount δ under the load. That deflection (or sag) can be calculated using a formula from the theory of strength of materials:

$$\delta = \frac{PL^3}{48EI} \tag{1.1}$$

Here L and I are, respectively, the length and the moment of inertia* of the cross-sectional area of the beam, and E is the modulus of elasticity (which is a measure of stiffness) of the material of which the beam is made. This formula is one of the standard formulas in structural mechanics. However, while its use is quite broad, and while it is of reasonable accuracy for most practical problems of interest to the structural engineer, there are certain restrictions on the applicability of this formula. One is that the deflection (or sag) of the beam must be small when compared to the thickness of the beam. Therefore, if we were walking across a simple wooden bridge of the kind we see on many a country path, the amount that the bridge deflects or gives has to be very small compared to the thickness of the bridge if Eq. (1.1) is to

*The moment of inertia is a property of the cross-sectional area of the beam that represents a resistance to rotation. The product EI is the stiffness or resistance of the beam to its being bent. See also the discussion following Eq. (6.10).

apply. This means that if d is the thickness of the simple beam, then

$$\frac{\delta}{d} = \frac{PL^3}{48EId} << 1 \tag{1.2}$$

This is a restriction that we can certainly verify, in numerical terms, both for a beam already built and for a proposed beam yet to be designed.

Another important restriction on the use of this formula is that the beam is assumed to be elastic. This means that the behavior of the material of which the beam is made is such that when a load is removed the beam will return to its undeformed, unstressed state. Further, it will return to this initial state by the same path by which it got to its stressed state. The assumption of elasticity in the beam model is incorporated into Eqs. (1.1) and (1.2) by the constant value of the modulus E and by the restriction of the inequality of Eq. (1.2). We can also check that the limitations of linear elasticity are not violated. This may be done by calculation or by actual physical measurement.

In this discussion of beams, we have twice noted points where we can check formulas to verify that their underlying assumptions have not been violated. There are other checks. For example, in order for Eq. (1.1) to make sense, the physical dimensions on both sides of the equation must be the same. That is, the deflection δ on the left-hand side of Eq. (1.1) will be calculated in inches or centimeters or some other measure of length. The net dimension of the load, the length, the modulus, and the moment of inertia—as represented in PL^3/EI—must also be measured in terms of length. This provides another type of check on the model, one that is based on the principle that all terms in an equation must have the same dimensions.

We are not interested in developing the theory of beams here, since that is a subject more advanced than and somewhat removed from our present concerns. However, this formula is so well known to engineers and scientists that it is a useful point of departure for a discussion of some of the ideas involved in defining a model, and of some of the assumptions or restrictions that may underlie the model's use. Thus, the kinds of checks that we have noted for this simple beam model will be amplified throughout this book, as will the ideas involved in generating such models.

It is also appropriate to note here that the philosophical idea that we are expounding is not original with us. In fact, it is a hallmark of the scientific method. The underlying philosophical structure has been discussed by philosophers of science, by mathematicians and physicists, by people interested in modeling, such as operations researchers and

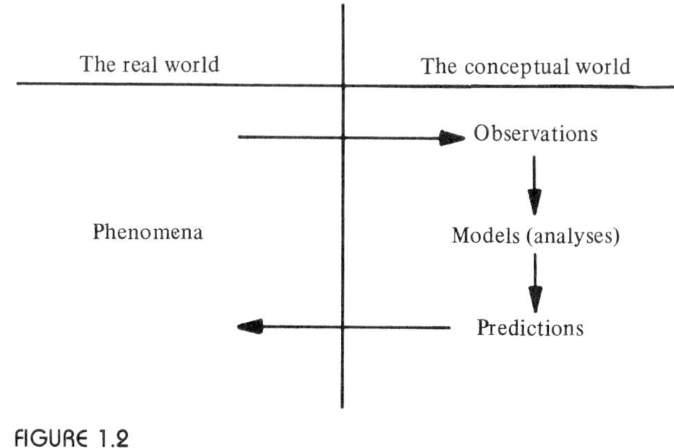

FIGURE 1.2
A scheme of the method of science

mechanicians, and by many other practitioners. What we are discussing is really a part of the scientific method. One way of characterizing the scientific method is displayed in Figure 1.2. In this figure we have identified a "real world" and a "conceptual world." The external world is what we call reality, wherein we observe and take note of various phenomena that occur in nature, whether natural or of human origin. The conceptual world is the world of the intellect. This is the world within our minds that we live with, talk about, and think about when trying to understand what goes on in the other (the real) world. We have divided this conceptual world into three stages: observations, models, and predictions.

The observation portion relates to our perceiving the real world, and to our integrating in some fashion an understanding of what is going on in the real world so that we can deal with it. These observations may be direct, as with using one of our senses; or indirect, in which case we may use elaborate scientific equipment. The point is that this portion of the scientific method is devoted to the gathering of data, so that we can inform ourselves about the world that we are perceiving. This information retrieval process can be done most sensibly and most efficiently if we have some idea of what we are looking for; that is, if we have some model that we can use as a guide to tell us what to expect. Thus, even though we have not yet discussed the modeling part of the scientific method, we can see already that the modeling part will help guide the observation part of the scientific method.

The second major component of the scientific method in the development of the conceptual world is the development of models to analyze situations. Here we are referring to the application of models, leaving aside for the moment the question of their generation, to help us understand the observations we have made. This understanding comes about by using the models as guides for observations, as predictors for future observation, and as tests of the validity and consistency of our observations. The model is of central importance for informing our observations as well as for making predictions. In other words, the modeling process is intricately entwined with the observation and prediction processes. The observations are in themselves a guide to the model, and the model is in turn a guide to observation.

From an engineering viewpoint we can go much further in dealing with the role of models because they provide the (analytical) basis for design. That is, after models have been proven and their ranges of validity tested and understood, they are then used to provide information about situations that have not yet been observed because they do not yet exist. In more concrete terms, for example, the well-developed theory of structures allows us to analyze a projected building before it is built, and we can do this with great confidence if we are careful to remember the assumptions inherent in the model. Thus, in the engineering design process, a model plays a crucial role not only in observation and pure prediction, but also in a kind of modified prediction that allows us to assess consequences in advance. It is not a question, as in the pure prediction model, of predicting something and then observing whether the predictions will be verified. Rather, it is a process of predicting with such confidence that we can spend resources of time, imagination, and money to design and build something, knowing in advance that the outcome will be successful.

The final stage in a pure scientific method is that of prediction, and as we have already seen, prediction is intricately tied into the observation and modeling processes and is in fact informed by both. We have also seen that the pure prediction process is modified, or sublimated if you will, into the design process for building something new. Again, in the pure scientific model, the role of the model is clear: The model provides the basis for informed prediction, and the predictions are then followed by observation to test the validity of the model. If the observations of subsequent behavior agree with the predictions based on the model, we can consider our model to have been verified. If not, then further work is required in order to generate a model that gives more accurate predictions.

In this business there are obviously many subtleties and philosophical points that can be the subject of endless debate; it is not our intent to compress into just a few paragraphs the entire subject matter of the philosophy of science. However, this backdrop is important because it sets the stage for what we are about. Our concern in this book is with the modeling process, and in particular with how models are developed, how they can be checked for internal consistency, and how they can be used as predictors in the design context as well as in the pure scientific context. Toward this end we have written this book in two parts: The first part deals with foundations, the second with applications.

The first part of the book examines mathematical tools of modeling. We will introduce in sequential chapters the concepts of dimensional analysis, scaling, and the numerical and analytical approximations of functions. Under the heading dimensional analysis we will discuss physical dimensions and units, the process of choosing variables, and the concepts of developing dimensionless groups of variables. In Chapter 3 we will deal with the concepts of scaling, size and shape, how scaling reflects on the boundary conditions of a problem, and the numerical consequences of choosing a particular scale. Then we will discuss how we can develop approximations and how we can assess the validity of approximations in numerical terms. Here we will be interested in series expansion, evaluation by series of trigonometric and other complicated functions, curve fitting, analysis of errors, and some of the concepts involved in model validation and prediction.

The second part of the book deals with applications. Here we will present a sequence of models that have relevance in the physical and natural sciences, as well as in some aspects of engineering. We will show how the foundation material is used to develop models and assess their validity. We will use models as predictors. We also demonstrate how information can be derived from models—in many instances without obtaining what is classically termed a solution—and we will see how such information is of interest to us, and in what ways it can be made more interesting by a different presentation. Some of the models that we will consider—particularly the linear and nonlinear models of oscillations of the pendulum—are very classical in nature. The pendulum model is a problem from which much information can be gained, and that information will highlight an all-important distinction between linear and nonlinear models. The linear version of the pendulum model in turn sets the stage for discussion of periodic motion in a variety of different physical applications. These include such diverse examples as the vibrations of a building, the response of an automobile suspension, the vibrations of a particle in a magnetic field, and the response of an acoustic resonator. We also discuss some other important physical ideas such as impedance and resonance.

We then discuss a variety of different models taken from other scientific and engineering fields. First we investigate a "car-following" model in which we look at the consequences of different assumptions regarding people's behavior in monitoring the safe distance between themselves and the car in front of them. These models, both linear and nonlinear, reflect some of the difficulties that occur when trying to assess human behavior accurately. However, we will see that analytical descriptions are available and that the models based on how an individual driver of a car monitors the behavior of the preceding car can be successfully integrated into overall models of highway capacity and traffic speed.

Chapter 8 deals with exponential models. Here we discuss growth rates, exponential behavior, and estimates of future populations of objects as well as people. We will discuss ways of approximating these estimates and will also discuss the interactions of the assumptions and estimates in the field of human population. We will demonstrate exponential behavior in such diverse fields as economics and finance, transportation, and the attrition of opposing armies at war.

This last model is one of the classic models of operations research, and so it quite naturally sets the stage for Chapter 9, in which we will discuss optimization with calculus and linear programming, the transportation problem, network analysis, and related topics. Finally, in Chapter 10 we will return to physics and look at optical and acoustical diffraction, where the crucial elements are the relative size or scale of objects as compared to the length of light or other waves.

In each of the extended examples and applications, we will use the basic principles developed in the first part of the book to illustrate how models are constructed and validated, and to provide estimates of the reasonableness of answers obtained from the models. Also, we will integrate empirical observations available in the literature to substantiate our models. The second section is, in a very real sense, self-contained. That is, it has been written so that the material therein can be read directly, with recourse to the first part of the book required only as specific techniques or items of information are needed.

PART A

FOUNDATIONS

2

DIMENSIONAL ANALYSIS

This chapter is the first of three devoted to detailing *tools* of mathematical modeling, that is, techniques that will be useful in developing and checking mathematical abstractions of physical (or otherwise "real") problems. The technique to be outlined here is called *dimensional analysis* and it has been effectively defined as follows:

> *Dimensional analysis is a method by which we deduce information about a phenomenon from the single premise that the phenomenon can be described by a dimensionally correct equation among certain variables.*
>
> H. L. Langhaar

That is, perhaps, a somewhat vague definition and, as Langhaar goes on to say, "The generality of the method is both its strength and its weakness." We shall endeavor in the balance of this chapter to give more substance to this definition; to discuss the roles played by dimensions, units, and dimensionless groups; and to show why dimensional analysis is both important and useful.

DIMENSIONS AND UNITS

We shall, for a significant portion of this book, be dealing with *physical quantities* which are *concepts*, such as time, velocity, and force, that can be measured or expressed *numerically*. That is, we see physical quantities as consisting of both a *concept* (or an abstraction) and a *numerical measure*. To partially describe a football field, for example, we could say that it is a playing field that is 100 yards long. The concept invoked here is length or distance, and the numerical measure is 100 (yards), or equivalently, 300 (feet). The numerical measure implies a comparison with some sort of *standard*, the standard being used to facilitate communication so that two quantities can be compared without their being placed physically side by side. The standard used for numerical measurement and comparison is itself arbitrary; it is chosen largely for convenience of reproduction.

It is useful (as well as conventional) to treat some physical quantities as *fundamental* or *primary*; others are then considered to be *derived* quantities. For example, in mechanical problems the fundamental quantities may be taken either as force, length, and time or as mass, length, and time. (We shall generally take mass, length, and time as the primary mechanical variables, leaving force to be derived from Newton's law of motion.) That a quantity is taken as fundamental rather than derived means only that it can be assigned a standard of measurement independent of that chosen for the other fundamental quantities. For a given problem, of course, there must be a sufficient number of primary quantities so that each derived quantity can be expressed in terms of these primary quantities.

For example, length and time are generally chosen as primary quantities, and velocity is a derived quantity expressed as length per unit time. However, we could choose time and velocity as primary quantities, in which case the derived quantities of length, area, and volume would be obtained as velocity · time, (velocity · time)^2, and (velocity · time)^3, respectively. We ought to note that while the primary quantities are chosen arbitrarily, the derived quantities are chosen to satisfy physical laws or some definition.

The word *dimension* is brought into play to relate a derived quantity to the fundamental quantities that have been selected for a particular analysis. For example, if the fundamental quantities are chosen as mass, length, and time, then we can say that the dimensions of area are (length)^2, the dimensions of mass density are mass/(length)^3, and the dimensions of force are mass · length/(time)^2. If we let $\mathscr{M}$, $\mathscr{L}$, and $\mathscr{T}$ stand for mass, length, and time, respectively, we can write these results symbolically as

$$
\begin{aligned}
[A = \text{area}] &= \mathscr{L}^2 \\
[\rho = \text{density}] &= \mathscr{M}/\mathscr{L}^3 \\
[F = \text{force}] &= \mathscr{M}\mathscr{L}/\mathscr{T}^2
\end{aligned}
\tag{2.1}
$$

where the brackets [ ] should be read as "the dimensions of."

The *units* of a quantity are the numerical dimensions of that quantity as expressed in terms of some physical standard. A unit is, therefore, some arbitrary multiple or fraction of a given standard. When measuring length, for example, the most widely accepted international standard is the meter (m). However, length can be measured in units of centimeters (1 cm = 1 m/100) or units of feet (0.3048 m). The magnitude of the numerical measure depends on the unit chosen, and this dependence may often dictate the choice of units so as to facilitate numerical computation. The football field, again, can be said to have a length of 100 yards, 300 feet, or 9144 centimeters.

We can now relate dimensions to units by noting that the dimensions of a quantity allow us to compute the numerical measure of that quantity in different systems of units. As we have just seen, the length of a football field can be expressed in terms of yards, feet, meters, or centimeters (to name but a few), but whatever the units, these measures all have the dimensions of length. Further, because the dimensions of a quantity are the same, we can write relations between the units. For example,

$$\begin{aligned} 1 \text{ foot} &= 30.48 \text{ centimeters} \\ 1 \text{ centimeter} &= 0.000006214 \text{ miles} \\ 1 \text{ hour} &= 60 \text{ minutes} = 3600 \text{ seconds} \end{aligned}$$

This equality of units for a given dimension allows us to change units in a straightforward algebraic manner. For a speed of 88 m/sec we may calculate the following equivalent:

$$\begin{aligned} 88. \frac{\text{meters}}{\text{second}} &= 88. \frac{\text{meters}}{\text{second}} \times \frac{3600. \text{ seconds}}{\text{hour}} \times \frac{100. \text{ centimeters}}{\text{meter}} \\ &\times \frac{0.000006214 \text{ miles}}{\text{centimeter}} = 196.9 \frac{\text{miles}}{\text{hour}} \end{aligned}$$

Note that each of the multipliers in this equation has the effective value of unity because of the equivalence of the units, which is, in turn, a consequence of using terms with the same dimensions. We shall discuss *systems of units* later in this chapter.

Before turning to the motivation for and the process of dimensional analysis, we must introduce the important notion of *dimensional homogeneity*. This notion shows up in two forms, one being a definition of rational equations, the other showing again the relation between units and dimensions. A *rational equation* is defined as an equation in which all terms have the same dimensions, so that the equation as a whole is dimensionally homogeneous. This means, simply, that we cannot add length to area in the same equation—although we can add together length in meters and length in feet.

Further, an equation that is rationally homogeneous does not depend on what the fundamental units of measurement are. For example, we shall see in Chapter 5 that the period (or cycle time) of a pendulum can be expressed in terms of the pendulum length l and the gravitational constant g as

$$T_o = 2\pi\sqrt{l/g} \tag{2.2}$$

This equation is valid, and dimensionally homogeneous, irrespective of the units used to measure length and time. However, if we replace the gravitational constant by its value in the English system of units, $g = 32.174$ ft/sec², we then get a result

$$T_o(\text{sec}) = \left(\frac{2\pi}{5.672}\right)\sqrt{l} \tag{2.3}$$

that is valid *only* if the length is given in units of feet. Equation (2.3) is not dimensionally homogeneous.

DIMENSIONAL ANALYSIS—MOTIVATION

Dimensional analysis is an analytical tool that is an outgrowth of attempts to perform costly and extended experiments in a more organized and efficient fashion. The basic idea was to see if variables could be grouped together in a meaningful way so as to reduce the number of trial runs or the number of measurements needed to get relevant data from an experiment. In the process, of course, the output of the experiment is obtained more compactly, thus reducing the number of graphs or charts that might be plotted, and so perhaps also clarifying the nature of physical processes being observed.

As an example, let us consider the experiment depicted in Figure 2.1, which shows the drag force being measured on a scale model submarine. The drag force is the force exerted by the water flowing by the submarine, the force that retards the motion of the submarine. The variables in this problem include the drag force F_D, the mass density of the water ρ, the velocity V of the submarine relative to the water, a characteristic dimension d of the submarine (which could be its length, circumference, or other typical indicator of size), and the viscosity of the water μ. The viscosity is a parameter that represents a coefficient of proportionality between the viscous friction stress (force/area) on a moving object and the spatial gradient of the velocity of that moving object. For this problem, we can take mass, length, and time as the fundamental or primary quantities. The derived quantities will be the velocity V, the density ρ, the viscosity μ, the submarine dimension d,

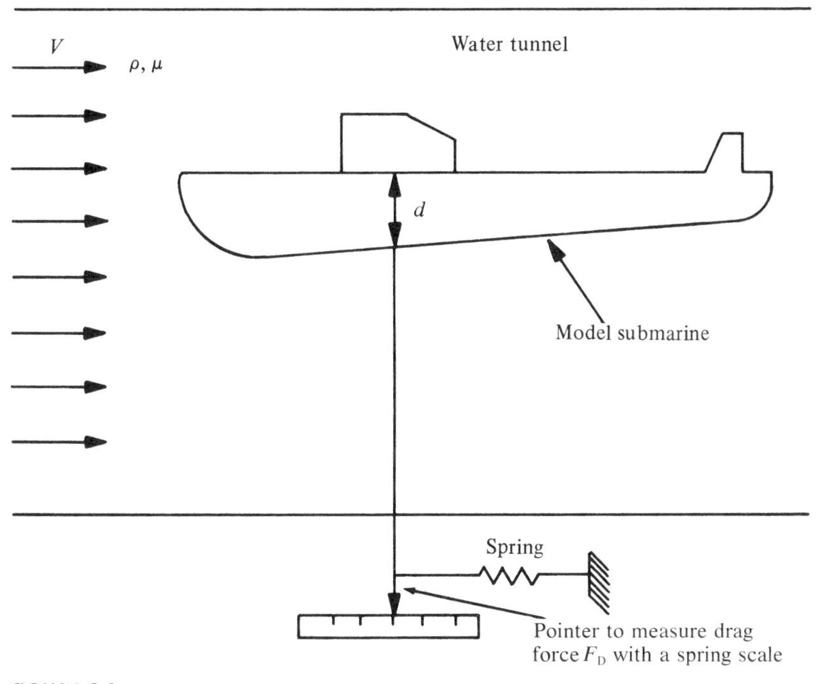

FIGURE 2.1
Experiment to measure drag on model submarines

and the drag force F_D. These variables are listed in Table 2.1 together with their dimensions expressed in terms of the fundamental dimensions.* We now visualize the experiments required to find out how the drag force varies with velocity, for example. But since there are five variables, we could wind up with a set of curves like those shown in

Table 2.1
Dimensions for Submarine Experiment

Variable (Symbol)	Dimensions
Relative velocity (V)	$\mathscr{L}\mathscr{T}^{-1}$
Characteristic dimension (d)	$\mathscr{L}$
Density (ρ)	$\mathscr{M}\mathscr{L}^{-3}$
Viscosity (μ)	$\mathscr{M}\mathscr{L}^{-1}\mathscr{T}^{-1}$
Drag force (F_D)	$\mathscr{M}\mathscr{L}\mathscr{T}^{-2}$

*The dimensions of viscosity are, by definition of that parameter, (force/area)/(velocity/distance). Since force is mass · length · (time)$^{-2}$, we have that

$$[\mu] = (\mathscr{M}\mathscr{L}\mathscr{T}^{-2}/\mathscr{L}^2)/(\mathscr{L}/\mathscr{L}\mathscr{T}) = \mathscr{M}\mathscr{L}^{-1}\mathscr{T}^{-1}$$

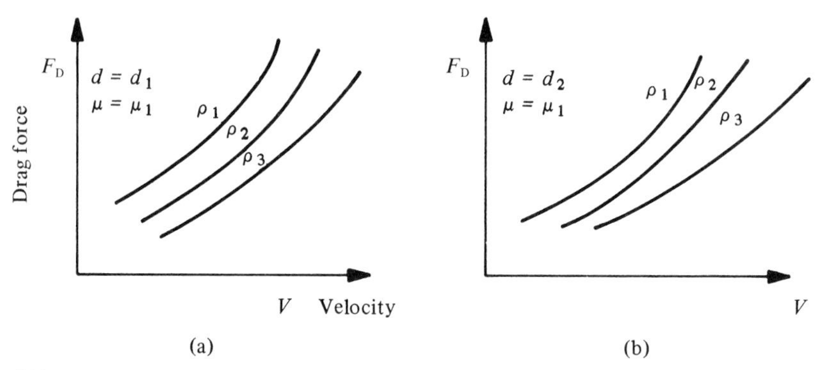

FIGURE 2.2
Schematic curves for submarine experiments

Figure 2.2. We see here that because of the large number of physical variables involved in the problem, and there are five, we will have to plot and examine many such curves to find the effect of varying the density, the submarine dimension, and the viscosity. If we were to look at three different values of each of ρ, d, and μ, we would have nine (9) different charts, each containing three (3) curves! This significant accumulation of data for a comparatively simple experiment provides a graphic illustration of the motivation for dimensional analysis that we referred to previously. We shall show in fact that this problem can be reduced to consideration of only two *dimensionless groups* that are related by a single curve. Also, the simplification of data gathering and the identification of crucial dimensionless groups can be extremely valuable in the design of an experiment.

There is additional reason to use dimensional analysis. It not only provides a means to simplify experiments; dimensional analysis also provides a useful tool for guiding the thinking about analytical problems and for identifying small or large effects in problems. In the submarine problem, for example, we shall find that the two dimensionless groups represent, first, the ratio of the drag force to the force of the fluid flowing against the projected frontal area of the model submarine; and second, the ratio of inertia forces to friction forces. The latter ratio, called the Reynolds number, is small for smooth (laminar) flow and large for rough (turbulent) flow. Hence, if we were interested only in turbulent flow, we could either design our experiment or conduct our analysis for cases where the Reynolds number is large enough for us to be sure that turbulent flow is being modeled. Such forethought often allows the analyst to drop terms from an equation, so as to simplify the solution, if the proper dimensional considerations have been addressed.

DIMENSIONAL ANALYSIS—THE PROCESS

The point of dimensional analysis is to ensure that we are dealing in any given problem with the right dimensions, whether expressed in terms of the proper number of correctly dimensioned variables or whether expressed in the appropriate dimensionless groups. In another perspective, dimensional analysis is that process whereby we guarantee that all terms in an equation have the same dimensions. Again, we can't add length to area. Further, while consistent units need to be used for numerical calculations, consistent dimensions are required for logical consistency. Thus we may measure speed in miles per hour or kilometers per second, but we want to be sure we are dealing with speed and not, say, distance.

How do we ensure this dimensional consistency? We do it first by checking the dimensions of all derived quantities to see that they are properly represented in terms of the primary quantities and their dimensions. The next step is the identification of the proper *dimensionless groups* of variables, that is, those ratios and products of the prob lem parameters and variables that are themselves dimensionless. Ther are several methods of exposing these dimensionless groups; we sha develop two of them, the classical *Pi theorem of Buckingham*, and th *basic method*.

Buckingham's Pi theorem has two parts. The first states that an equation that is dimensionally homogeneous can be reduced to a set (dimensionless products or ratios, so this part of the theorem is akin t an existence statement. The second part of the Pi theorem is as follow The number of independent dimensionless groups of variables need to correlate the variables in a given process is equal to $n - m$, where is the number of derived variables involved and m is the number primary variables or dimensions. Thus, in the submarine experime where we have derived five variables (see Table 2.1) and three fu damental dimensions, we should expect to find two dimensionle groups.

The Pi theorem can be applied as follows. Let $A_1, A_2, \ldots, A_n$ be t n derived variables of the process being analyzed. Choose m of th such that these m derived variables contain among them all of primary dimensions. Then form the dimensionless groups with the maining $n - m$ variables as follows: If A_1, A_2, A_3 are the chosen deri variables, form the groups as

$$
\begin{aligned}
\Pi_1 &= A_1^{a_1} A_2^{b_1} A_3^{c_1} A_4 \\
\Pi_2 &= A_1^{a_2} A_2^{b_2} A_3^{c_2} A_5 \qquad (\\
\Pi_{n-m} &= A_1^{a_{n-m}} A_2^{b_{n-m}} A_3^{c_{n-m}} A_n
\end{aligned}
$$

where the a_i, b_i, c_i are chosen so as to make the groups Π_i dimensionless. For the submarine model, for example, choose the velocity V, the characteristic dimension d, and the density ρ from among the derived variables, and let us permute them with the remaining two variables (the viscosity μ and the drag force F_D) to get two dimensionless groups:

$$\begin{aligned} \Pi_1 &= V^{a_1} d^{b_1} \rho^{c_1} \mu \\ \Pi_2 &= V^{a_2} d^{b_2} \rho^{c_2} F_D \end{aligned} \tag{2.5}$$

In terms of the primary dimensions, these groups are

$$\begin{aligned} \Pi_1 &= \left(\frac{\mathscr{L}}{\mathscr{T}}\right)^{a_1} \mathscr{L}^{b_1} \left(\frac{\mathscr{M}}{\mathscr{L}^3}\right)^{c_1} \frac{\mathscr{M}}{\mathscr{L}\mathscr{T}} \\ \Pi_2 &= \left(\frac{\mathscr{L}}{\mathscr{T}}\right)^{a_2} \mathscr{L}^{b_2} \left(\frac{\mathscr{M}}{\mathscr{L}^3}\right)^{c_2} \frac{\mathscr{M}\mathscr{L}}{\mathscr{T}^2} \end{aligned} \tag{2.6}$$

For Π_1 and Π_2 to be dimensionless, the exponents for each of the primary dimensions must vanish. Thus, for length, time, and mass, respectively, we must have

$$\begin{aligned} \mathscr{L}&: \ a_1 + b_1 - 3c_1 - 1 = 0 \\ \mathscr{T}&: \ -a_1 - 1 = 0 \\ \mathscr{M}&: \ c_1 + 1 = 0 \end{aligned}$$

and

$$\begin{aligned} \mathscr{L}&: \ a_2 + b_2 - 3c_2 + 1 = 0 \\ \mathscr{T}&: \ -a_2 - 2 = 0 \\ \mathscr{M}&: \ c_2 + 1 = 0 \end{aligned}$$

so that $a_1 = b_1 = c_1 = -1$, $a_2 = b_2 = -2$, and $c_2 = -1$. Then the two dimensionless groups for the submarine model experiment are

$$\begin{aligned} \Pi_1 &= \frac{\mu}{\rho V d} \\ \Pi_2 &= \frac{F_D}{\rho V^2 d^2} \end{aligned} \tag{2.7}$$

These are the groups we anticipated for this problem, reflecting in the first instance the ratio of viscous to inertial forces and, in the second group, the ratio of the drag force to the force exerted by moving water

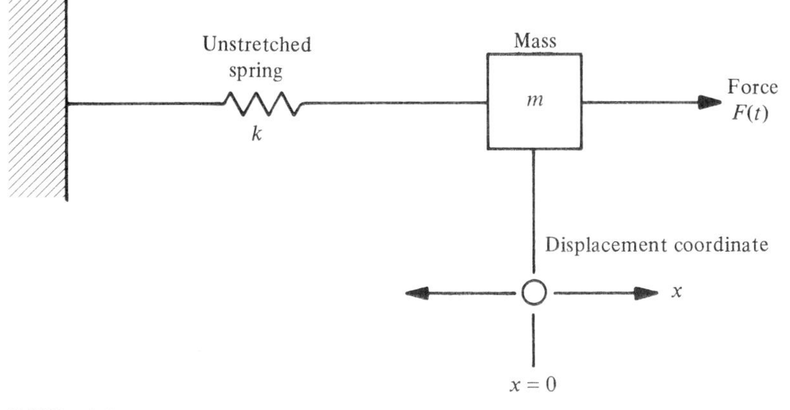

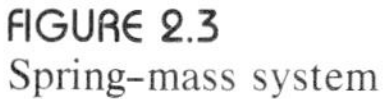
FIGURE 2.3
Spring-mass system

on the projected frontal area of the submarine. The relationship of these two groups to each other lets us describe the behavior of the submarine.

Another problem that we shall deal with in great detail in Chapter 6 is the forced vibration of a spring-mass system, also called a simple oscillator, for which a schematic is displayed in Figure 2.3. For our present interest, we are satisfied with the listing of derived variables and dimensions given in Table 2.2. As the chosen derived variables, we use the maximum displacement A, the mass m, and the spring stiffness k. Then, with the usual three primary variables ($\mathscr{L}$, $\mathscr{M}$, and $\mathscr{T}$), we can expect to form two dimensionless groups from

$$\begin{aligned} \Pi_1 &= A^{a_1} m^{b_1} k^{c_1} F_0 \\ \Pi_2 &= A^{a_2} m^{b_2} k^{c_2}\, \omega \end{aligned} \tag{2.8}$$

By following the same procedure as we did for the submarine model,

Table 2.2
Variables and Dimensions for Forced Oscillator

Variable (Symbol)	Dimension
Maximum displacement of mass (A)	$\mathscr{L}$
Mass (m)	$\mathscr{M}$
Spring stiffness (k)	$\mathscr{M}\mathscr{T}^{-2}$
Magnitude of applied force (F_0)	$\mathscr{M}\mathscr{L}\mathscr{T}^{-2}$
Frequency of applied force (ω)	$\mathscr{T}^{-1}$

we can calculate that $a_1 = c_1 = -1, b_1 = 0, a_2 = 0, b_2 = \frac{1}{2}$, and $c_2 = -\frac{1}{2}$, and so we obtain the groups

$$\Pi_1 = \frac{F_0}{kA}$$
$$\Pi_2 = \omega\sqrt{m/k} \equiv \frac{\omega}{\omega_0} \tag{2.9}$$

In writing the second of Eqs. (2.9) we have identified $\omega_0 = \sqrt{k/m}$ as the *natural frequency* of the spring-mass system; see Eq. (6.7). Thus the second dimensionless group is the ratio of the frequency of the applied force to the natural frequency of the system being forced. The first group is the ratio of the applied force to the force in the spring, since the spring force must be equal to the spring stiffness times its extension (or compression).

These dimensionless groups can also be treated as dimensionless variables that relate the system response to its basic properties. The first group in Eqs. (2.9), for example, can be seen as the reciprocal of the ratio $A/(F_0/k)$. The denominator of this ratio, being the magnitude of the applied force divided by the spring stiffness, is just the static displacement of the spring. The aforementioned ratio, for the dynamic problem, tells us whether the total response is (depending on its magnitude) smaller than, comparable to, or larger than the static response. Thus we can gauge the dynamic effect in the system. Similarly, the magnitude of the second ratio of Eqs. (2.9), ω/ω_0, when compared to unity, tells us whether the system is being forced at a frequency below, near, or above its natural frequency.

It is instructive to repeat this calculation with a variation; that is, we will choose a different set of the derived variables as the set around which the others are permuted. This time we shall construct the groups around the force amplitude F_0, the frequency ω, and the mass m. Thus we seek the modified groups

$$\Pi_1' = F_0^{a_1}\omega^{b_1}m^{c_1}k$$
$$\Pi_2' = F_0^{a_2}\omega^{b_2}m^{c_2}A$$

By the same techniques as before, we can calculate that $a_1 = 0$, $b_1 = -2$, $c_1 = -1$, and $a_2 = -1$, $b_2 = 2$, $c_2 = 1$. Then it follows that

$$\Pi_1' = \frac{k}{m\omega^2} = \frac{\omega_0^2}{\omega^2} = \left(\frac{1}{\Pi_2}\right)^2$$
$$\Pi_2' = \frac{\omega^2 mA}{F_0} = \frac{\omega^2}{\omega_0^2}\left(\frac{kA}{F_0}\right) = (\Pi_2)^2\left(\frac{1}{\Pi_1}\right) = \frac{1}{\Pi_1\Pi_1'}$$

where Π_1 and Π_2 are defined in Eqs. (2.9). It is clear that Π_1' and Π_2', although dimensionless, are not the same as Π_1 and Π_2, although they are equally clearly related. In fact, what we have demonstrated by this exercise is that the dimensionless groups determined in any one calculation are in one sense unique, but they can appear in different forms, as evidenced here by the variety of ways in which we can write Π_1' and Π_2'.

The *basic method* of dimensional analysis is less formal, less structured. Once the significant variables are identified, we construct a *functional equation* in which all the variables are contained, and for which all the dimensions are noted. The proper dimensionless groups are then identified by a thoughtful elimination of dimensions. Let us illustrate this with a couple of examples.

First we consider the free fall of a body in a vacuum. We know that the fall must be related to the gravitational acceleration g and the height h from which the body is released, and we are particularly interested in how these parameters affect the velocity V of the falling body. We start with a functional equation that expresses in general terms the dependence of V on g and h:

$$V = V(g,h) \tag{2.10}$$

where

$$\begin{aligned} [V] &= \mathscr{L}\mathscr{T}^{-1} \\ [g] &= \mathscr{L}\mathscr{T}^{-2} \\ [h] &= \mathscr{L} \end{aligned} \tag{2.11}$$

Note that the discussion of time appears only in the velocity and the gravitational constant. Thus, if we divide the left-hand side of Eq. (2.10) by $\sqrt{g}$, we can eliminate the time dimension altogether, and get a new functional relationship:

$$\frac{V}{\sqrt{g}} = V_1(h) \tag{2.12}$$

Now we have reduced the number of variables in the argument of the functional relationship [on the right-hand sides of Eqs. (2.10) and (2.12)] by one.

We repeat this process to make the left-hand side dimensionless in length as well as time. We must then have

$$\frac{V}{\sqrt{gh}} = \text{constant} \tag{2.13}$$

Equation (2.13) is a familiar result from physics, and the constant is $\sqrt{2}$. Before proceeding to another example, we can point out here very quickly why the velocity of a body in free fall in a vacuum is independent of the body's mass. Had we incorporated the mass into our thinking, we would have written Eq. (2.10) as

$$V = V(g, h, m) \tag{2.14}$$

Now, there is no way that we can render this functional equation dimensionless with respect to the mass because mass appears in none of the other variables, so we can assume that the free fall velocity is indeed independent of the mass.

As another example of the basic method, let us develop the appropriate groups for the spring–mass system. Here we start with a functional equation for the maximum displacement of the oscillator or, equivalently, of the mass at the end of a spring:

$$A = A(m, k, F_0, \omega) \tag{2.15}$$

We now make all the terms dimensionless with respect to length, say, by combining F_0 with them to eliminate the length. This results in the form

$$\frac{A}{F_0} = A_1(m,\ k,\ \omega) \tag{2.16}$$

where

$$[A/F_0] = \mathcal{T}^2\mathcal{M}^{-1} \tag{2.17}$$

and the dimensions of m, k, and ω are as listed in Table 2.2.

We now start out to eliminate the dimension of mass on the left-hand side, but we see immediately that the dimensions of A/F_0 are exactly those of $1/k$. Hence, eliminating the mass dimension by multiplying by the stiffness k, we find that

$$\frac{kA}{F_0} = A_2\left(\frac{m}{k},\ \omega\right) \tag{2.18}$$

where m and k are combined in A_2 so that

$$[m/k] = \mathcal{T}^2 \tag{2.19}$$

Hence, noting the dimensions of the forcing frequency ω, we have

$$\frac{kA}{F_0} = A_3\left(\frac{m}{k}\ \omega^2\right)$$

or, in view of the definition of the natural frequency ω_0 that is given in the second of Eqs. (2.9),

$$\frac{kA}{F_0} = A_3\left(\frac{\omega^2}{\omega_0^2}\right) \tag{2.20}$$

Thus we find here the same two dimensionless groups, kA/F_0 and ω/ω_0, that we found through application of the Pi theorem of Buckingham.

As a final example of the basic method, we consider the revolution of two bodies in space, one about the other, the revolving motion being caused by the mutual gravitational attraction. We wish to find a (dimensionless) functional relation for the period of the motion. The derived variables and their dimensions are listed in Table 2.3. Then the period can be expressed as

$$T_R = T_R(M_1,\ M_2,\ R) \tag{2.21}$$

But we see immediately that our list of variables cannot be complete, for there is no variable with a dimension of time on the right-hand side of Eq. (2.21). In contrast to a different conclusion that we drew from Eq. (2.14), for the falling body, we conclude here that since we are looking for a variable whose dimension is time itself, we must have left something out; and, indeed, we have, for we have no representation of the gravitational force of attraction between the two bodies. We can account for this force either by including it directly, or by incorporating the gravitational constant G from Newton's law of gravitational attraction, where

$$[G] = \mathscr{L}^3\mathscr{M}^{-1}\mathscr{T}^{-2} \tag{2.22}$$

Table 2.3
Variables and Dimensions for Revolving Bodies

Variable (Symbol)	Dimensions
Mass of first body (M_1)	$\mathscr{M}$
Mass of second body (M_2)	$\mathscr{M}$
Separation distance (R)	$\mathscr{L}$
Period of revolution (T_R)	$\mathscr{T}$

Thus, the correct functional equation for the period is

$$T_R = T_R(M_1, M_2, R, G) \tag{2.23}$$

Now, applying the basic method to eliminate the dimension of time first, we have

$$T_R G^{1/2} = T_{R1}(M_1, M_2, R) \tag{2.24}$$

where

$$[T_R G^{1/2}] = \mathscr{L}^{3/2}\mathscr{M}^{-1/2} \tag{2.25}$$

It is now a simple matter to eliminate both the dimensions of mass and length on the left-hand side of Eq. (2.24) in one step:

$$\frac{T_R G^{1/2} M_2^{1/2}}{R^{3/2}} = T_{R2}\left(\frac{M_1}{M_2}\right) \tag{2.26}$$

This example illustrates the problem of starting with an incomplete set of variables. There was, moreover, no obvious reason to include the gravitational constant G until it became clear that we were headed down a wrong path. Then the constant was included, to rectify an incomplete analysis. We could have argued, as we did with the benefit of hindsight, that somehow the effect of the attractive gravitational force had to be incorporated. This argument, as with the deletion of the mass in the free fall problem, depends heavily on insight and judgment whose origins may have little to do with the problem at hand.

Both of these examples do serve to point out that dimensional analysis is not a tool that can be applied in an obvious or mechanical way. Some thought is required, since both the inclusion of extraneous variables and the deletion of essential variables can lead to errors and inconsistencies. Our use of dimensional analysis as a formal tool will not be extensive in the remainder of this book. The underlying ideas about variables and their dimensions are, however, very important to what follows.

UNITS

We have already noted that units are numerical measures derived from standards, and so are either arbitrarily fractions or multiples of these standards. In the United States, the system of reference units in common use is the *English* system of units, in which length is referenced in feet, force in pounds force, time in seconds, and mass in slugs. The *pound force* is defined as that force which gives an acceleration of 32.1740 ft/sec^2 to a mass that is 1/2.2046 part of a certain piece of platinum known as the *standard kilogram.* Although units and dimensions ought not be confused, it is worth observing that in the English system the fundamental reference quantities (foot, pound, second) are based on the primary variables of length, force, and time.

A new reference system of units is gradually being adopted in this country. That system, known as the SI* system, is a modern version of the metric system and it is based on the physicists' mks system. In the SI system the fundamental reference standards are the meter, the kilogram, and the second, for the primary variables of length, mass, and time, respectively. Force is a derived variable in the SI system, and it is measured in newtons.

In Table 2.4 we have summarized some of the salient features of the SI system, and some of the commonly used conversion factors from

Table 2.4
Systems of Units and Conversion Factors

Reference units	English system	SI system
Length	Foot	Meter (m)
Time	Second	Second (sec)
Mass	Slug, pound mass	Kilogram (kg)
Force	Pound force	Newton (N)

Multiply number of	by	to get
Feet	0.3048	Meters
Inches	2.540	Centimeters
Miles	1.609	Kilometers
Pound force	4.448	Newtons
Slugs	14.59	Kilograms
Pound mass	0.454	Kilograms
Miles per hour	0.447	Meters per second

Numerical factors in SI system	Prefix name (symbol)
10^3	kilo (k)
10^{-2}	centi (c)
10^{-3}	milli (m)
10^{-6}	micro (μ)

*For *Système International.*

the English system to the SI units. It is also worth stressing in the same way that all terms in a rational equation must have the same dimensions, numerical calculations must use a consistent set of units throughout.

SUMMARY

In this chapter we have presented an approach to problem formulation and experimental design. That approach is called dimensional analysis. The steps in this method are (1) identification of relevant physical variables and constants needed to describe all aspects of the problem or experiment; (2) selection of a system of primary variables and dimensions; and (3) the development, through one of several methods (e.g., the basic method, Buckingham's Pi theorem), of the proper collection of dimensionless groups or variables. These dimensionless groups can be used to estimate the importance of various effects, to buttress our physical insight and understanding, and to simplify experiment design, data gathering, and numerical computations.

Again we note that although we will not make extensive use of the method for deriving dimensionless groups, we will both explicitly and implicitly make use of the notions of dimensionless variables and groups. We shall see these applications in scaling effects, in solutions to problems, and in our various model formulations. Further, we shall always keep in mind the differences between dimensions and units, as well as the importance of maintaining dimensional homogeneity and consistency of units.

Problems

2.1 The discharge Q of a fluid through a capillary tube is thought to depend on the pressure drop per unit length $\Delta p/l$, the diameter d, and the viscosity μ. Show that only one dimensionless group can be formed, from which it follows that $Q = (\text{const})(d^4/\mu)(\Delta p/l)$. The discharge is also called the volume flow rate.

2.2 Find the four dimensionless groups for the pressure drop Δp for flow of a viscous fluid inside a rough pipe. As a measure of roughness, use the average variation of pipe radius e. Other variables are the fluid velocity V, the fluid mass density ρ, the pipe diameter d, the pipe length l, and the viscosity μ.

2.3 Consider a string of length l connected to a fixed point at one end and to a rock of mass m at the other. The rock is whirling in a circle at constant velocity v. Use dimensional analysis to show that the force in the string is determined from the dimensionless group

$$\frac{Fl}{mv^2} = \text{constant}$$

2.4 The speed of sound in a gas depends on the pressure p and the mass density ρ. Use dimensional analysis to show that the speed of sound is proportional to the square root of the pressure and is inversely proportional to the square root of the density.

2.5 The speed of sound in an elastic solid depends on the modulus of elasticity E and on the mass density. Use dimensional analysis to show how the speed depends on the modulus and the density.

2.6 The Weber number (in fluid mechanics) is a dimensionless number involving the surface tension σ (which has dimensions of force/length), fluid velocity v, mass density ρ, and a characteristic length l. Use dimensional analysis to find that number.

2.7 A pendulum swings in a vacuum. Use dimensional analysis to find the single dimensionless group that relates the period of the pendulum T_0 to its length l and to the gravitational acceleration g. Why isn't the mass of the pendulum involved?

2.8 A pendulum swings in a viscous fluid. In addition to the usual pendulum variables, we must now consider the fluid viscosity μ, the density ρ, the diameter d of the pendulum, and the angle of the pendulum swing. Decide how many groups are appropriate and find them.

2.9 The compliance (the deflection per unit load) for a beam of square cross section of $d \times d$ depends on the length l of the beam, on the width and thickness d, and on the modulus of elasticity E (which has dimensions of force per unit area) of the material of which the beam is made. Show by dimensional analysis that

$$CEd = \text{function of} \left(\frac{l}{d}\right)$$

2.10 In an experiment it is found that for a number of beams a plot of $\log_{10}(CEd)$ against $\log_{10}(l/d)$ has a slope of 3 and an intercept on the $\log_{10}(CEd)$ scale of -0.60. Show that the deflection under a load P can be given in terms of the moment of inertia of the cross section ($I = d^4/12$) as

$$\text{deflection} = \text{load} \cdot \text{compliance} = P \cdot C = \frac{Pl^3}{48EI}$$

3

SCALING

In this chapter we shall continue to deal with issues of dimensions, but we will focus now on the aspects of *scale*, that is, on the aspects of relative size. Size, both absolute and relative, is a very important aspect of modeling because in the broadest possible way it influences both form and function. The relations of objects (both animate and inanimate) to their environments, shape and function in nature, the design and technique of experiments, the representation of discrete data by smooth curves are all concepts that are significantly influenced—if not controlled outright—by scaling. We even find reference to scaling in literature, as in this quotation describing Gulliver's treatment in the land of Lilliput:

> *His Majesty's Ministers, finding that Gulliver's stature exceeded theirs in the proportion of twelve to one, concluded from the similarity of their bodies that his must contain at least 1728 of theirs, and must needs be rationed accordingly.*
>
> Jonathan Swift

We will thus devote this chapter to discussions of size and shape, and of size and function, to outlining the effects of scaling on boundary conditions, and to looking at the consequences of choosing a particular scale.

SIZE AND SHAPE

Consider first some simple geometrical scaling ideas. We show in Figure 3.1 two cubes of different size. The units with which the various dimensions might be measured are not important here, so we simply

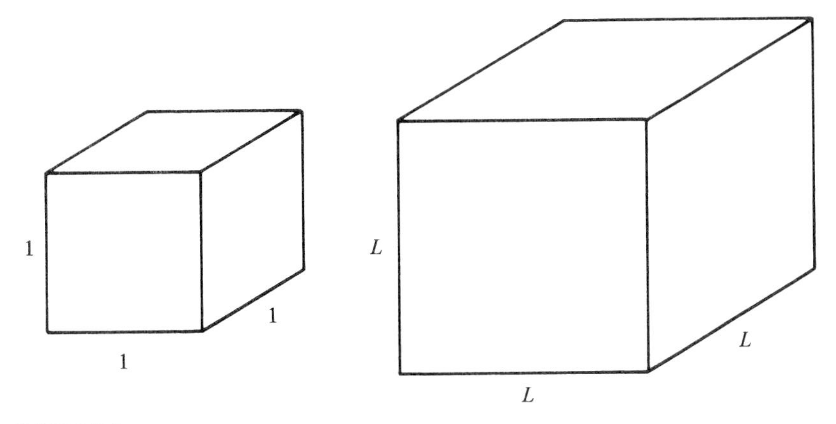

FIGURE 3.1
Two cubes of different size

take the lengths of the sides of the cubes shown as 1 and L, respectively. The total surface area and the volume of the cubes are respectively 6 and 1, and $6L^2$ and L^3. We see then that as the ratio of the lengths of a side of a cube changes by a factor of $L/1 = L$, the total surface area of the cube changes as L^2, and the volume of the cube changes as L^3. Thus, doubling the length of the side of a cube increases the surface area by a factor of four and the volume by a factor of eight.

We can also look at this geometrical scaling in another way. Consider two geometrically similar bodies, on each surface of which we measure corresponding distances. Then, if the distance between two points on one body is in the ratio of $1/L$ to the distance between the corresponding points on the second body, the ratios of the corresponding surface areas and volumes of the two bodies are, respectively, $1/L^2$ and $1/L^3$. We may illustrate this by considering two points marked on the surface of a balloon of radius 1. Obviously the arc between two points on the surface of a sphere is proportional to the radius of that sphere, as may be seen by blowing up the balloon from a radius 1 to a radius R. The subtended angle between the points on the arc of the balloon will not change as the radius is increased, so the arc length change must be directly proportional to the change in the radius. Since surface area and volume for a sphere depend on, respectively, the square and the cube of the radius, the rest of the argument follows in a straightforward fashion.

As an aside, we note that simple geometrical scaling arguments can be used to illustrate the ideas of linear proportionality and of linearity and nonlinearity. Consider the four glasses depicted in Figure 3.2. For the cylindrical glasses with the same radius r, the volume of fluid in

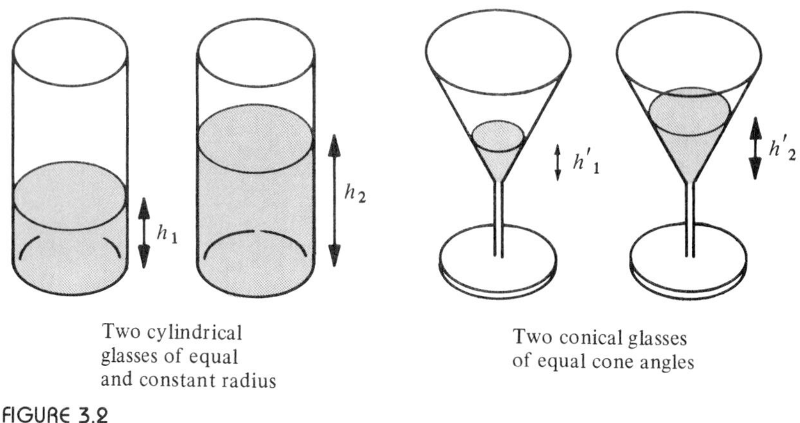

FIGURE 3.2
Glasses illustrating linearity

each can be written as $\pi r^2 h_1$ and $\pi r^2 h_2$, respectively. The total fluid volume is then

$$V = \pi r^2 h_1 + \pi r^2 h_2 = \pi r^2 (h_1 + h_2) \tag{3.1}$$

We see that the volume is proportional to the first power of the height of the fluid column, and so volume is here *linearly proportional* to the height. Further, we can get the total volume by simply adding the two heights; therefore, the volume is a *linear* function of the height. The volume is obviously not a linear function of the radius, though.

In the case of the conical glasses, since the radius depends on the height also, we obviously cannot write anything like the right-hand side of Eq. (3.1). For a cone of semivertex angle α, in fact, the volume is

$$V = \frac{1}{3}\pi \frac{h^3}{\tan^2 \alpha} \tag{3.2}$$

Hence, the total volume of fluid in the two conical glasses of Figure 3.2 is

$$V = \frac{1}{3}\pi \frac{(h_1')^3}{\tan^2 \alpha} + \frac{1}{3}\pi \frac{(h_2')^3}{\tan^2 \alpha} \neq \frac{1}{3}\pi \frac{(h_1' + h_2')^3}{\tan^2 \alpha} \tag{3.3}$$

That is, the relationship between volume and height for the cone is *nonlinear*, and we cannot find the total volume as a superposition of the two heights h_1' and h_2'. Thus, while for a cylinder the volume of fluid is linearly proportional to the height of the fluid column and scales accordingly, the volume of fluid in a cone scales as the cube of the height of the fluid column and is not a linear function of that height.

If we go a bit beyond purely geometrical considerations, we will find scaling even more intriguing. We shall be able to think about why there is an upper limit to the weight of a bird or to the height of a tree, about why the height of a jumping mammal and the maximum speed of a running animal are both independent of the animal size, about how what we hear depends on the size of the eardrum, and about how the quality of a person's voice depends on the size of the vocal cords. Let us begin to look at this aspect of scaling by considering some empirical data and seeing if we can make any sense of it.

The first set of data is derived from examination of medieval churches and cathedrals in England. We show in Figures 3.3 and 3.4 some drawings of the naves* of cathedrals and a plot of the heights of cathedral naves against church lengths. The church lengths can be taken as indicators of church size. Thus we see from Figure 3.4 that the larger churches have naves that are higher in absolute terms than those of smaller churches, yet in *relative terms* the nave heights are smaller in the larger buildings. Further, although we've not given the data to

FIGURE 3.3
Drawings of cathedral naves. (a) The nave of St. Albans, the oldest Romanesque cathedral in England, is relatively low in proportion to length. (b) The nave of Canterbury Cathedral, an example of the late Gothic style called perpendicular, is relatively high. (Used by permission of Professor S. J. Gould, Harvard University.)

*The nave is the principal longitudinal area of a church, extending from the entrance at one end to the chancel or altar area at the other.

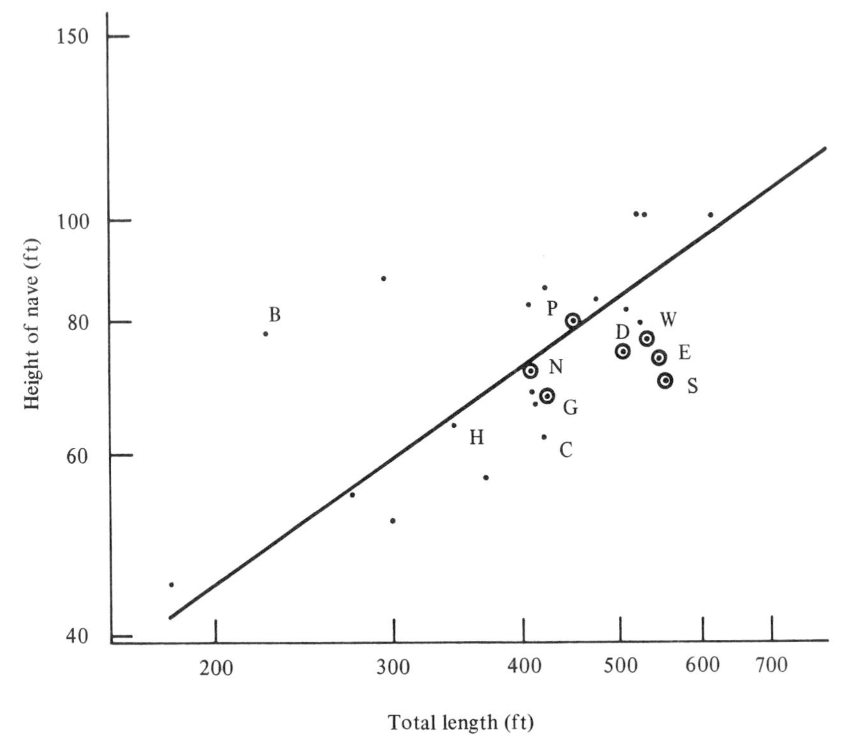

FIGURE 3.4
Plot of cathedral nave height against cathedral length. As churches get longer, the naves rise proportionately less. In this graph, the churches of England are indicated as follows: B, Earls Barton; C, Chichester; D, Durham; E, Ely; G, Gloucester; H, Hereford; N, Norwich; P, Peterborough; S, St. Albans; W, Winchester. Circled dots indicate Romanesque churches. (Used by permission of Professor S. J. Gould, Harvard University.)

buttress this assertion, the bigger churches also tend to have narrower naves. Why don't the nave height and width increase proportionately as the church is made larger? The answer has to do with the scale of size and shape as well as with the building technology of the times. In particular, the answers lie in the scaling of surfaces and the volumes they enclose.

The scaling that is relevant here is that of the change of the enclosed area of a cathedral as it is made longer (or larger) by lengthening its perimeter. In buildings of constant shape, the periphery of the outside wall increases as $(\text{length})^1$ while the enclosed area increases as $(\text{length})^2$. Thus there is an increasing problem of the penetration of exterior light and ventilation. (Remember, these churches and cathedrals were built before electric lights and air conditioning!) However, this problem can be circumvented by changing the shape, for example by incorporating transepts,* or by maintaining one dimension to be relatively constant. If the width is kept constant, then the enclosed area will also increase as $(\text{length})^1$ with the perimeter, and the resulting building will appear to be relatively narrow.

To enlarge the width as well as the length not only increases the area inside, and therefore the lighting and ventilation problems; it also requires a roof that can span a larger width. That is, if the width is enlarged, a greater surface is required to cover the resulting enlarged volume in the church. Since roofs were basically built on stone vaults or arches, the width of the church was very important because it was impossible to build long stone arches. This also relates to the problem of nave height, for it is the nave walls that support the outward thrust of the roof vaults, even in the presence of flying buttresses. Thus, higher nave walls required additional thickness not only for the support of the additional self-weight, but also for the support of the vaults at the more flexible tops of higher walls. Therefore, in sum, the width and height of cathedral naves had to be scaled down as overall cathedral size (or length) was increased lest there be insoluble problems of illumination, ventilation, and structural safety.

SIZE AND FUNCTION

The second set of empirical data arises from an investigation into the aerodynamics of birds in flight; it is displayed in Figure 3.5. It appears from this plot that the data seem to peter out at weights of 35–40 lb, so we are prompted to ask two questions: Can the general form of the data be explained by dimensional reasoning? Is there an upper limit to the weight of a bird? The answers are affirmative (at least for flying birds, since birds like the ostrich can far exceed that nominal weight). For soaring birds, for example, we note that the lift forces that sustain the bird in the air must be proportional to the wing area, or to the dimension of $(\text{length})^2$. The loading on the wing is simply the weight of the bird, which must be proportional to the volume of the bird; therefore the loading is of dimension of $(\text{length})^3$. Hence the wing loading plotted as the ordinate in Figure 3.5, which is the loading per unit area on the wing, is simply proportional to $(\text{length})^3/(\text{length})^2$ or to (length). This means that

$$\text{wing loading} \sim \text{length}$$
$$\text{weight} \sim (\text{length})^3$$

*The transept of a cathedral is a section laid out perpendicular to the longitudinal axes near the location of the altar. The result of adding a transept is a cruciform shape, though the origin may be of other than religious design.

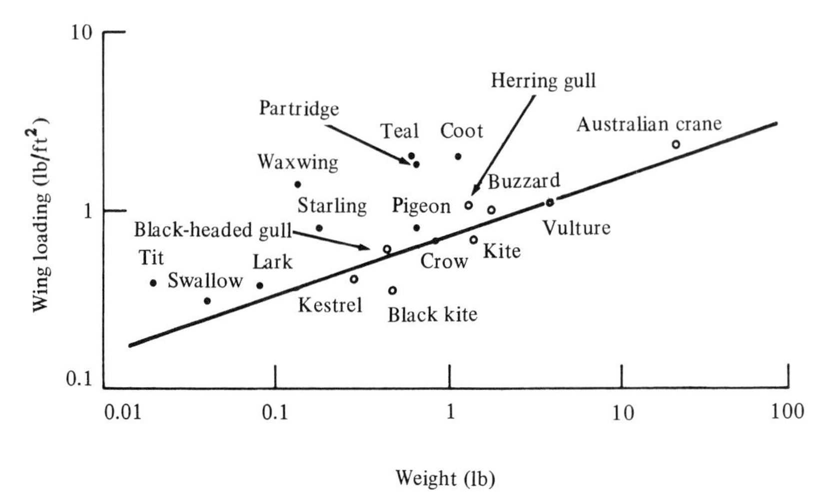

FIGURE 3.5
Wing loading versus weight for birds in flight. Open circles denote the birds which regularly soar, filled circles those which flap their wings. (Reprinted from Theodore von Karman, *Aerodynamics*, copyright © 1954 by Cornell University. Used by permission of the publisher, Cornell University Press.)

We would then expect a line on a logarithmic plot to have a 1:3 slope to confirm this postulate about flying; and, indeed, we see that this is a good fit to the data.

The question of the upper bound to the weight is harder to answer. We will restrict our attention to the simplified aerodynamics of hovering flight, and we shall examine the relationship between the power *required* to sustain hovering and the power *available* to sustain hovering. The power required can be estimated from the momentum of the downward jet of air generated by the flapping wings: The time rate of change of the jet momentum must be equal to the total lift force on the bird, which is just the weight. The mass of air moved through the jet per unit time can be expressed in terms of the density of the air ρ, the area of the jet A (which is also the wing surface area), and the velocity v of the air mass, as

$$\text{mass/time} = \rho A v$$

The momentum per unit time of the air jet is just $v \cdot (\text{mass/time})$, which must equal the weight W of the bird:

$$W = \rho A v^2 \tag{3.4}$$

Since the weight is proportional to $(\text{length})^3$ of the bird and the area is proportional to $(\text{length})^2$, the velocity of the air jet must be such that

$$v \sim \mathscr{L}^{1/2} \tag{3.5}$$

The power required to sustain the jet is equal to the kinetic energy of the mass of air in the jet per unit time. Thus,

$$\text{power required} \sim \tfrac{1}{2}\rho A v \cdot v^2$$

and so in view of Eq. (3.5) we have

$$\text{power required} \sim \mathscr{L}^2\mathscr{L}^{1/2} \cdot \mathscr{L} = \mathscr{L}^{7/2} \qquad (3.6)$$

This is a result, incidentally, that is valid for forward flight as well as hovering, and it can be confirmed by more complete (and complex) aerodynamic arguments.

The power available can be estimated from a heat loss argument, from an estimate of the rate at which the heart supplies oxygen, or from limits on the maximum stresses in bones and muscle. The heat loss argument is simple, if not altogether compelling. Muscles, which turn chemical energy into mechanical energy, operate at an efficiency of approximately 25%. The excess energy must be dissipated as heat through the body's surface area, so the temperature differential must decrease at a rate proportional to $\mathscr{L}^2$. Hence the power available must also be proportional to $\mathscr{L}^2$ to prevent overheating of the body. The oxygen supply argument reduces to consideration of the volume of blood delivered by the heart per unit time, a volumetric rate that is proportional to the cross-sectional area of the blood vessels. Hence, again we find that power is proportional to $\mathscr{L}^2$ on the basis that the deliverable power is proportional to the oxygen flow, which is proportional to the rate of blood delivery.

The final argument comes from equating the work done by a contracting muscle to the change in the kinetic energy of the limb it moves. Therefore,

$$\text{muscle force} \cdot \text{contraction} = \text{limb mass} \cdot v^2$$

where v is the velocity of the limb in question. Since the muscle force is limited to the maximum tensile strength of the muscles and tendons, it must be proportional to $\mathscr{L}^2$ as representative of the cross-sectional area of the muscles and tendons. Since the contraction is proportional to $\mathscr{L}$, and mass to $\mathscr{L}^3$, we find that limb velocity is independent of length; that is, velocity is independent of size.* If this be true, the time taken for a muscle to contract is proportional to length/v or simply to $\mathscr{L}$. Then the power output of the muscle is

$$\text{power} = \frac{\text{muscle force} \cdot \text{contraction}}{\text{time}} \sim \frac{\mathscr{L}^2 \cdot \mathscr{L}}{\mathscr{L}}$$

and so again we find that the power available is proportional to $\mathscr{L}^2$.

*We have also just shown that the maximum speed at which animals run is independent of size!

Thus we have shown that the *power required* for flight is proportional to $\mathscr{L}^{7/2}$ while the *power available* for flying is proportional to $\mathscr{L}^2$. Since the required power increases with bird size much more rapidly than the available power, it is clear that an upper limit to flying size must indeed exist.

An area of human physiology where scaling is interesting and important is that of hearing and speech. Size, shape, and function are clearly intertwined in the eardrum and the vocal cords, as well as in the "voice box," or larynx, which contains the vocal cords. We are inclined to ask about scale effects in hearing because we know that while humans hear sounds in the range of 20 to 20,000 hertz,* dogs hear sounds that have frequency components up to 50,000 Hz, and bats up to 100,000 Hz. What role does size play in this difference? The answer depends in part on the configuration of the eardrum, and in particular on its radius.

The eardrum is only one part of the hearing apparatus, which, starting at the outer ear and ending at the organ of corti, converts the (mechanical) vibration of air particles that is sound into the electrical signals by which information is transmitted through the nervous system to the brain. The eardrum is the first pickup of the sound vibration, and it is important that the eardrum remain very stiff so as to accurately pick up and transmit to the hammer–anvil–stirrup portion of the ear even the high-frequency components of the sound.

In mechanical terms the eardrum is a stretched membrane. As such it has *natural frequencies* of vibration, below the lowest of which it reacts as an elastic spring. If the eardrum is to be very stiff, it should have a very high natural frequency.† The lowest natural frequency of a circular stretched membrane of radius r with a tensile force F per unit length of circumference is

$$f_{\text{membrane}} = \frac{2.40}{2\pi r}\sqrt{\frac{F}{\sigma}} \tag{3.7}$$

where σ is the mass per unit area of the membrane material. Since the dimensions of σ are $\mathscr{M}/\mathscr{L}^2$, we can see as a quick check that the units of the frequency are $(1/\mathscr{L})(\mathscr{M}\mathscr{L}/\mathscr{L}\mathscr{T}^2)^{1/2} \cdot (\mathscr{M}/\mathscr{L}^2)^{-1/2} = 1/\mathscr{T}$, or 1 per unit time, which is appropriate for frequency. In particular, though, note that the membrane frequency varies as (1/radius), so that for similar tensile forces F and surface densities σ, the fundamental ear-

*The hertz (abbreviated as Hz) is a unit of frequency equal to 1 cycle per second. We shall discuss this concept at greater length in Chapters 5 and 6.

†See Chapter 6, in particular the section "Forced Motion of a Linear Oscillator: Resonance and Impedance."

drum frequency varies inversely to the size of the eardrum. Thus, for smaller animals we would expect the range of hearing to extend into higher frequencies than would be the case for larger animals.

A similar situation obtains with regard to the vocal cords and the voice box. We know from many observations that men generally have lower-pitched voices than women, and we are accustomed to the low-pitched growl of bears and the high-pitched chirping of birds. So we are inclined to look for a relation between size and frequency of sound. The mechanism of speech is the forced vibration of the vocal cords as air is pushed past them in the voice box (the larynx) after it is expelled from the lungs. In order to develop and produce volume at low frequencies, the vocal cords must be able to vibrate at low frequencies and the larynx must be able to amplify the low-frequency tones produced by the vocal cords.

The vibrations of the vocal cords are very similar to those of piano strings, and the fundamental or lowest natural frequency of either is given by the formula

$$f_{\text{string}} = \frac{1}{2l}\sqrt{\frac{F}{\sigma'}} \tag{3.8}$$

where l is the length of the string, F the tensile force in the string, and σ' the mass per unit length of the string. We see in Eq. (3.8) a marked similarity to the membrane result of Eq. (3.7), and it is also easily seen that the frequency for the string has the proper dimensions. Further, we see that if the vocal cords are increased in length and in density, the natural frequency decreases. Hence we would expect that a larger animal with longer and more dense vocal cords will develop sounds with components of lower frequencies.

In the same way we can look at the natural frequency of an acoustic cavity as a model for the voice box. Such a cavity is called an acoustic resonator or a Helmholtz resonator and we treat it in some detail in Chapter 6. For our purposes, it is sufficient to examine the fundamental frequency of the cavity, which is [see Eq. (6.23)*]

$$f_{\text{cavity}} = \frac{c_0}{2\pi}\sqrt{\frac{A}{l'V_0}} \tag{3.9}$$

where A and l' are the area and length of the "neck" leading into an

*Equation (6.23) is given as a radian frequency ω, which is related to frequency f by the formula $\omega = 2\pi f$. Thus the units of ω are radians per unit time. See Eq. (6.8) and the related text.

acoustical cavity of volume V_0 that contains a gas in which sound waves travel at a speed c_0. Clearly, the larger the volume of the cavity, the lower its natural frequency, so it is easier to produce loud sounds at low frequencies. Thus we again find that we expect larger animals to have deeper voices.

SIZE AND LIMITS

In the preceding section on size and function we demonstrated that there must be an upper limit to the weight of flying animals. This limit developed because of the discrepancy between available power and required power for flying as bird dimensions increased. Such limits occur quite often and in one way or another they may control the size and shape of objects, the range of variables in an equation, or even the application of some physical models (or "laws," as they are commonly called) themselves.

These limits can be imposed in a variety of ways. Twenty years ago, for example, portable computers and calculators were not possible because the computer circuit elements then available dictated large and bulky devices. Today we can hold computers in the palms of our hands. The advances in computer technology that have made possible such drastic reductions in size, weight, and cost have greatly expanded the role of computers in our society.

Other examples of how size may have limits occur in the (mechanical) design of parts where the limits may be set by the interactions between parts, by materials technology, or by function. A simple illustration is the choice of the length of high-tension wire to be strung between towers. The length of wire between any two sections must be somewhat longer than the straight-line distance between the towers—but not too much longer. If the wire is too short, the contraction of the wire during cold weather can generate unacceptably large tensile stresses in the wire. If the wire is too long, it can sag too close to the ground, particularly in warm weather when the wire will experience thermal expansion. Similar situations occur when the dimensional contractions or expansions are two-dimensional (surfaces) or three-dimensional (volumes). A plate that is designed to cover a certain hole snugly, for example, can be expanded or contracted over its length and its breadth, so that the surface expansion is proportional to $(\text{length})^2$. For very close-fitting pipes and pipe joints, the changes in the cross-sectional area of the pipe or of a fitting ring must be carefully examined. In designing pipes and rings, therefore, care must be taken to match the coefficients of thermal expansion.

Volume expansions involve changes in length, breadth, and height. Thus, this type of expansion is proportional to $(\text{length})^3$. Rates of linear, surface, and volume expansion are expressed in terms of an appropriate expansion coefficient. We shall show that an expansion coefficient for surfaces can be derived from the coefficient of linear expansion in Chapter 4, where we will be discussing several types of approximations.

Scaling may shift limits or perhaps boundary conditions in certain situations. For example, suppose we are interested in finding an approximation for the hyperbolic sine function,

$$\sinh x = \tfrac{1}{2}(e^{x} - e^{-x}) \tag{3.10}$$

We know that for large values of x, the term e^{-x} will be much smaller than the term e^{x}. The problem is to find some appropriate criterion for throwing away, perhaps, the term e^{-x}. Our thinking on this matter will be enlightened if we introduce a *scale factor*, λ; then we shall look for values of x for which

$$\sinh \lambda x \cong \tfrac{1}{2}e^{\lambda x} \tag{3.11}$$

A plot of $\sinh x$ is shown in Figure 3.6. For values of x greater than 3, the second term on the right-hand side of Eq. (3.10), e^{-x}, does become very small (less than 4.98×10^{-2}) compared to e^{x} for $x = 3$, which is 20.09. Hence, we could for most purposes take $\sinh x \cong \frac{1}{2}e^{x}$. All we need to do here is decide for what value of x we are willing to accept the approximation $e^{2x} - 1 = e^{2x}$. However, if the scale factor λ is introduced, as in the approximation in Eq. (3.11), then the value of x for which that approximation is acceptable will obviously depend on λ. We have shown how λ affects both the exact and the approximate forms of $\sinh x$ in Figure 3.7. For $\lambda = 1$, we can use the approximate form for $x \geq 3$, while for $\lambda = 5$ the approximation is good for $x \geq 0.5$. Thus, by changing the value of λ we are in effect changing a boundary condition if we view that boundary as the point at which the approximation is valid.

Another interesting example of the interaction of scale and limits is Newtonian mechanics. In both our everyday lives and in normal applications of mechanics we are accustomed to taking the masses (or weights) of objects as constants. Thus we do not expect a box of candy to weigh any more while we are standing still, riding in an automobile at 88.50 km/hr (55 mph), or flying in an airplane at 965 km/hr (600 mph). Yet, according to the general theory of relativity, the mass m of an

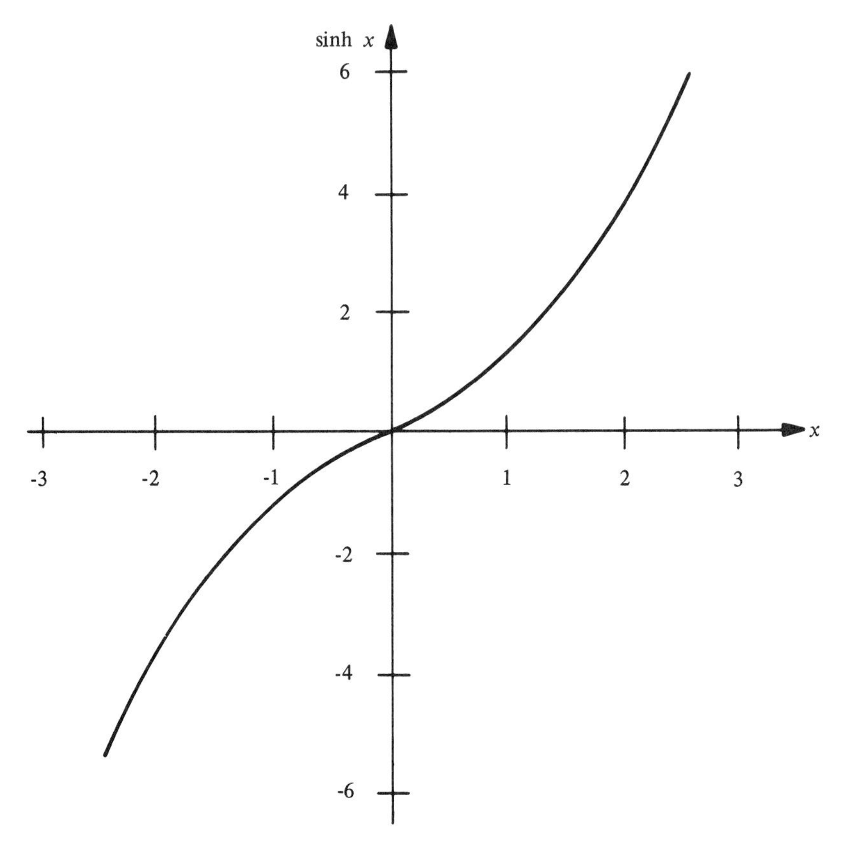

FIGURE 3.6
Hyperbolic sine function

object with rest mass m_0 is given by

$$m = \frac{m_0}{\sqrt{1 - \frac{v^2}{c^2}}}$$

where v is the speed with which the mass is moving and c is the speed of light (3×10^8 m/sec). For the airplane, the factor in the denominator of the relativistic mass formula becomes (noting that 965 km/hr = 268 m/sec)

$$\sqrt{1 - \frac{v^2}{c^2}} = \sqrt{1 - 7.98 \times 10^{-13}} \cong 1 - 3.99 \times 10^{-13} \cong 1.$$

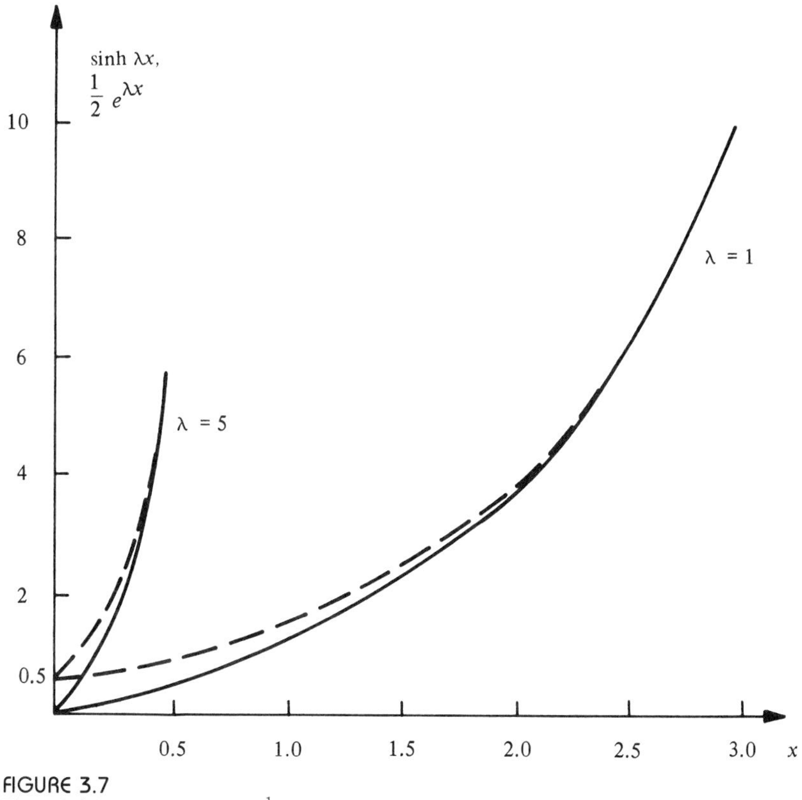

FIGURE 3.7
Plot of sinh λx (———) and $\frac{1}{2}e^{\lambda x}$ (- - -) for $\lambda = 1$ and $\lambda = 5$

Clearly, for our practical day-to-day existence, the relativistic effect is going to be entirely negligible. However, the point is clear that Newtonian mechanics is a good model only on a scale where all speeds are very much smaller than the speed of light. If the ratio v/c becomes significant, the (constant) rest mass m_0 cannot be taken as a representation of mass, and Newtonian mechanics must be replaced by relativistic mechanics.

In Chapter 10 we shall discuss diffraction patterns in great detail; that discussion will provide an extended example of limits controlling size and scale.

CONSEQUENCES OF CHOOSING A SCALE

The consequences of choosing a particular scale become very important in the design of an experiment, especially where data acquisition and reduction are concerned. With a careful choice of scale, errors can be reduced, time and money can be saved, and important details can be highlighted. Further, such consideration of scale obviously has implications for our observations of natural (''real life'') phenomena as well as for laboratory experiments. We shall illustrate some of these ideas with a few examples.

The choice of scale can affect the error factor during the acquisition of experimental data, as we see in the following simple experiment. We wish to measure the rotational inertia (moment of inertia) of a wheel; that is, we want to know how easily a wheel can be set into rotational motion. An experimental setup that allows us to measure the wheel's rotational inertia is depicted in Figure 3.8. In this experiment a falling weight is used to rotate the wheel. The weight provides a constant torque, which turns the wheel. The torque τ is related to the rotational inertia I by the formula

$$I = \frac{\tau}{\alpha} \tag{3.12}$$

where α is the angular acceleration of the wheel, measured in radians per second squared (rad/sec^2). The important parameter here for our discussion of scale and its consequences is α. The angular acceleration depends on the time t, the radius R of the wheel, and the speed v of any point (say, A) on the rim of the wheel as follows:

$$\alpha = \frac{v_f - v_0}{R(t_f - t_0)} \tag{3.13}$$

The subscripts f and 0 in Eq. (3.13) refer to readings of the speed and time taken at the beginning of the experiment (t_0, v_0) and at the end of the experiment (t_f, v_f). The time interval $t_f - t_0$ is the time scale here, and it will control the amount of error between the experimentally determined value of the rotational inertia of the wheel and the actual (or calculated) value.

The wheel is set into rotational motion by releasing a falling weight which is attached to the axis of the wheel by means of a string. As the weight falls faster and faster, the wheel rotates faster and faster. At the start of the experiment nothing is moving (that is, $t_0 = 0$, $v_0 = 0$). Thus, in this case the average value of the acceleration $\bar{\alpha}$ is

$$\bar{\alpha} = \frac{v_f}{Rt_f} \tag{3.14}$$

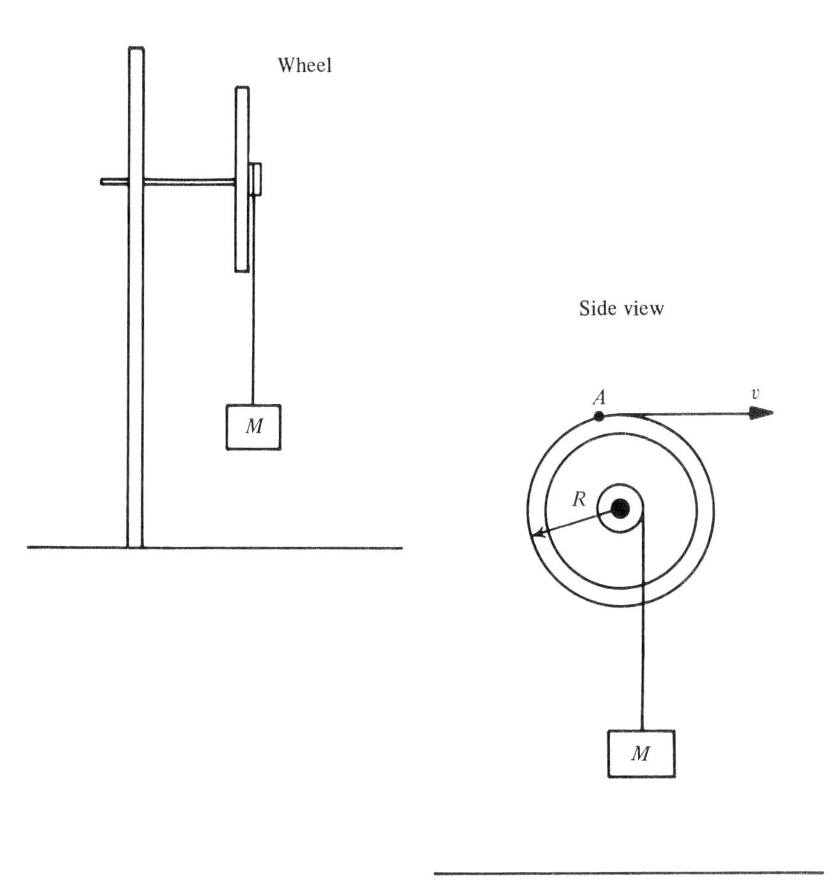

FIGURE 3.8
Apparatus to measure the rotational inertia of a wheel

Now, at the onset of motion static friction must be overcome so that the rate of change of the velocity (acceleration) is slowed during the first few seconds of motion. After a short while, the motion settles into a constant acceleration. For the sake of this discussion, can we say that the constant acceleration is reached after 2 sec? If we let the wheel rotate through three complete revolutions we find that it takes 4 sec and that the speed of a point on the rim is 1 m/sec. It takes 6 sec for 6 complete revolutions to occur, the point *A* on the rim of the wheel attaining a speed of 2.7 m/sec. These and additional data are shown in Table 3.1, along with the calculated value for $\bar{\alpha}$ when $R = 0.5$ m. We see here that the longer we allow the wheel to turn, the more constant our value of $\bar{\alpha}$ becomes. This stabilization occurs because the drag due to the initial static friction acts over an ever smaller percentage of the total time that the wheel is turning. Thus, in this case the longer the time scale over which our measurements are made, the more accurate our value for $\bar{\alpha}$ will be.

Suppose now that we have an electronic device that is not functioning properly and we are going to repair it. We suspect a certain component is malfunctioning, so we look at its output voltage on an oscilloscope. When this component is tested we expect to see a perfect sinusoid, say, although this is not true for all electronic components. This is our data source. At first glance the voltage looks like a "perfect" sine wave, which it ought to be in this case. But we notice that the horizontal time scale on the oscilloscope is set at a very high value, so we spread out the trace (and so shorten the time scale) by turning the appropriate knob. In doing so, we find that what we originally thought was a pure sine wave turns out otherwise (see Figure 3.9b). This confirms that the part in question is indeed malfunctioning. Had we not looked at the oscilloscope with the time scaled properly we might have come to the erroneous conclusion that the component was operating satisfactorily when, in fact, it was not, a fact illustrated by the poorly shaped sinusoid in Figure 3.9b.

In the area of acoustics it is often necessary to subject data to frequency analysis. Most sounds contain many different frequencies and this type of analysis measures the sound pressure for different parts of the frequency spectrum. Filters are used, filters that pass a specified range of frequencies and cut out all others. As the filters sweep through the frequency spectrum, the sound pressure level of each band of frequencies is measured and can be plotted as in Figure 3.10. Here we have plotted sweeps of three different types of filters. The octave-band

Table 3.1
Experimental Data and Resultant Values of Angular Acceleration

Number of revolutions	t_f (sec)	v_f (m/sec)	$\bar{\alpha}$ (rad/sec^2)
3	4	1	0.5
4	5	1.75	0.7
6	6	2.7	0.9
8	8	4	1
10	10	5	1

(a) Horizontal time scale of 0.5 sec/division

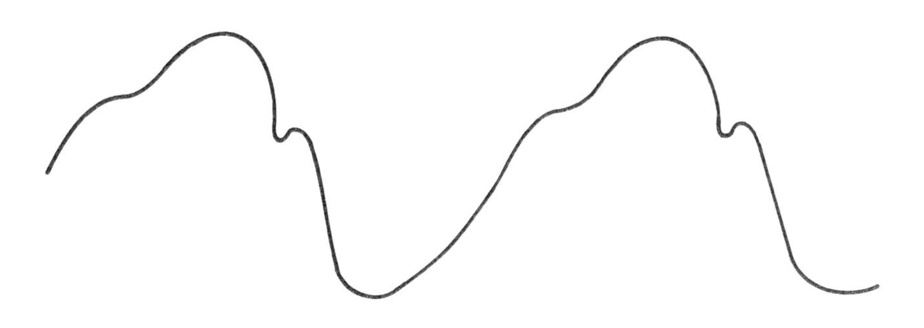

(b) Horizontal time scale of 0.5 msec/division

FIGURE 3.9
Sinusoidal oscilloscope traces

filter (Figure 3.10a) measures the sound pressure level (SPL) for all of the frequencies contained in an octave, that is, a band in which the lowest frequency is half that of the highest frequency. Sometimes octave-band analysis is not sufficiently selective. More detailed information is given by bands with the width of one third of an octave (Figure 3.10b), and still more detail can be obtained with "narrow-band" filters such as the one whose output is shown in Figure 3.10c. Narrow-band analysis is used to determine pure tones (single-frequency sounds), which often originate in machinery. These pure tones make some machinery, such as bandsaws and fans, sound very unpleasant. The pure tones may also indicate an unwanted source of vibration.

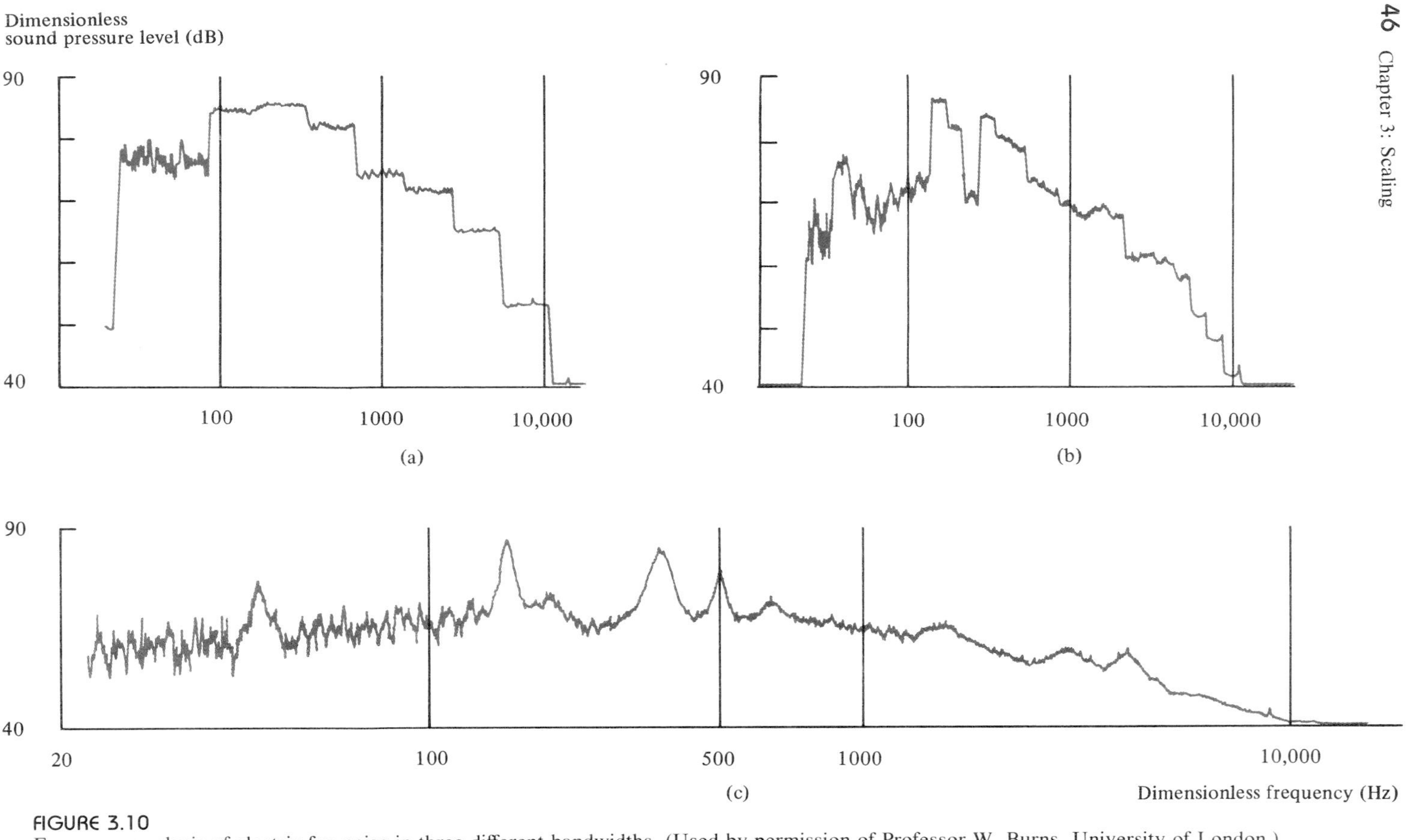

FIGURE 3.10
Frequency analysis of electric fan noise in three different bandwidths. (Used by permission of Professor W. Burns, University of London.)

Our third example involves the choice of coordinate scale when plotting data on a graph. Too contracted a scale may not show important details. To illustrate this we have redrawn Figure 3.7 (see Figure 3.11) with a different scale on the horizontal axis. Now we are unable to distinguish any differences between plots of sinh λx and $\frac{1}{2}e^{\lambda x}$. It looks as though they are the same for the whole range of x when, in fact, we know that for small values of x the curves are indeed somewhat different. With the scale of Figure 3.11 we have lost important information that occurs for $x < 3$. In the same way, had we redrawn Figure 3.7 on an overly long scale, we could have come to an opposite conclusion. To show this we have redrawn in Figure 3.12 the same plots of sinh λx and $\frac{1}{2}e^{\lambda x}$, and now these curves are exaggeratedly different. Thus in scaling data we must keep in mind what we are looking for, for example, in this case an approximation for large values of λx or for small values of λx.

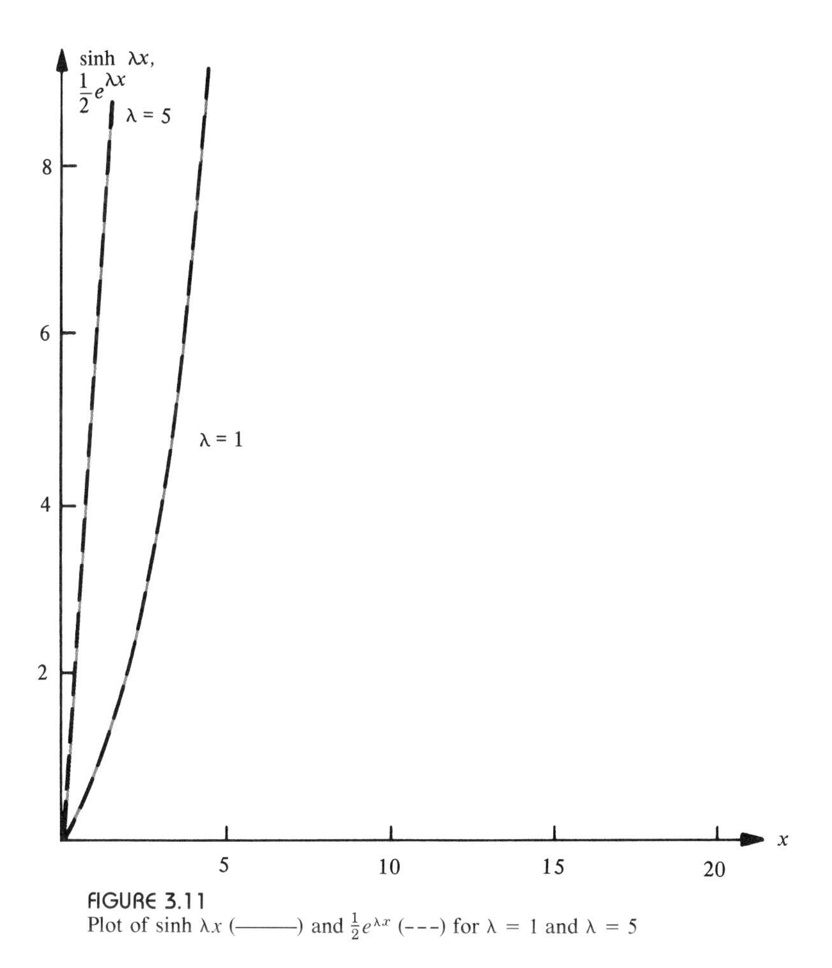

FIGURE 3.11
Plot of sinh λx (———) and $\frac{1}{2}e^{\lambda x}$ (- - -) for $\lambda = 1$ and $\lambda = 5$

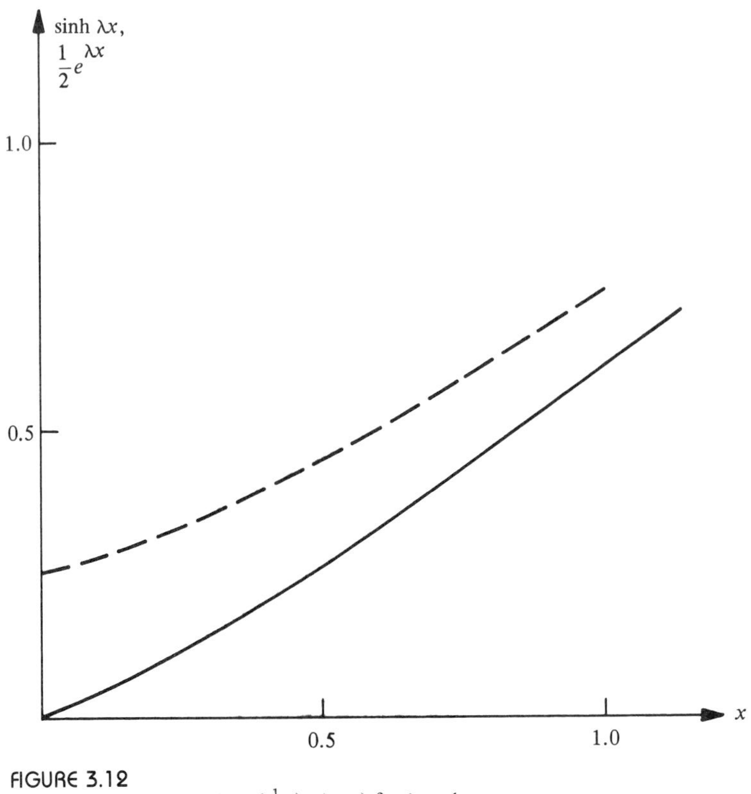

FIGURE 3.12
Plot of sinh λx (———) and $\frac{1}{2}e^{\lambda x}$ (- - -) for $\lambda = 1$

SUMMARY

In this chapter we have focused on the effects of scale in the continued discussion of issues involving dimensions. We have shown how scaling down the width and height of cathedral naves became important as the overall cathedral length increased. Such scaling solved the problems of illumination, ventilation, and structural safety. Similarly, size was shown to affect function in that certain size requirements must be met before birds can fly. Also, the speech and hearing frequency ranges of animals were shown to be determined in large part by the size of appropriate portions of their anatomy.

We have discussed the fact that experimental technique, particularly the design of the experiment and the acquisition and analysis of data, depends on the appropriate choice of a scale. One of the consequences of choosing an inappropriate scale is the loss of important information, which then leads to the drawing of erroneous conclusions. We illustrated this problem in several ways, including the determination of the rotational inertia of a wheel and the processing of acoustic data.

Problems

3.1 The velocity of blood in the aorta is related to the difference in pressure between the heart and the arteries. Find the relationship between the velocity of the blood and pressure. (*Hint*: Use the work-energy theorem in a way similar to that discussed on page 35.)

3.2 In *Gulliver's Travels*, a certain little long-legged bird, the stilt, weighed $4\frac{1}{2}$ oz and had legs 8 in. long. A flamingo is of similar shape and weighs 4 lb. Use scaling to show that a flamingo's legs should be about 20 in. long, as they actually are.

3.3 A certain cucumber was found to have cells that divided when they had grown to 1.5 times the volume of the "resting" cells. Normally, cells divide so that the ratio between surface and mass remains constant. Is this cucumber "normal"?

3.4 A scale model test is used to determine the natural frequencies of a beam supported on its two ends. Both the model and the prototype are made of the same material. If the lengths are scaled 5:1, how are the natural frequencies of the model and the prototype related? (*Hint*: First, find the dimensionless groups knowing that the frequency depends on mass density ρ, modulus of elasticity E, span length L, depth d, and cross-sectional area A. Then, the scaling law says $\Pi_{1p} = \Pi_{1m}$.)

3.5 A steel beam with a span of 20 cm is to be used to model a timber beam with a 3.6-m span. The length scale is designated as $n = l/l_{model}$.

(a) Verify that the dimensionless variable (group) containing the load P can be written as P/El^2, where E is the modulus of elasticity.

(b) Knowing that the timber beam must be able to carry a concentrated load of 9000 N at a point 1.5 m from the left end of the beam, calculate the load that should be used in the model to test the beam's ability to withstand that load.

3.6 Speculate on the reasons why a student performing the following experiment was unable to verify the period of oscillation (T) of a simple spring-mass system.

Experimental procedure: One end of a spring was attached to the end of an air track. The other end of the spring was attached to a mass m, as shown in the accompanying figure. The mass m moves

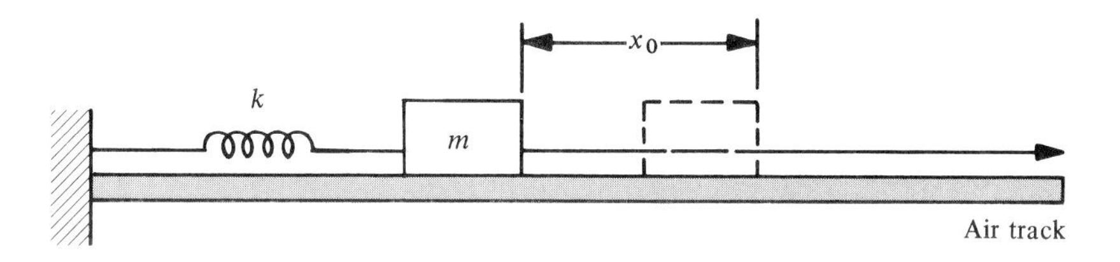

without friction on the air track. The student pulls the mass to the right a distance x_0 with a force F and then releases it. As the mass oscillates back and forth, the time for a complete oscillation (that is, the period) is measured for several periods in succession. These measurements are then compared to the calculated value using

$$T = 2\pi\sqrt{\frac{m}{k}}$$

The spring constant k and the mass m are known, and the timer is accurate to within $\pm 1\%$.

3.7 We want to use the approximation

$$\cosh \lambda x = \tfrac{1}{2}e^{\lambda x}$$

Find the range of x for which this approximation is acceptable when the scaling factor $\lambda = 1$, and for $\lambda = 6$. Plot both functions in each case.

3.8 Looking for the growth rate of a sample colony of bacteria, we record the data given in the accompanying table.
(a) Plot population as a function of time.
(b) Write an equation expressing the population as a function of time.

Time (min)	Population $(p) \times 10^5$
0	10
5	15
10	22
20	50
30	110
40	245
50	546
60	1,215
70	2,704
80	6,018
90	13,394
100	29,810

4

APPROXIMATION AND REASONABLENESS OF ANSWERS

In our continuing discussion of the mathematics of model building, we now look at approximation techniques and at ways to test and validate models. Approximation techniques often serve to simplify models, as in Chapter 5 where the linear pendulum model is discussed as an approximation (and quite a good one) to the complete nonlinear pendulum model. At the same time, however, we must be cautious, since approximation introduces "error" into a model. We will, therefore, discuss approximations using Taylor series, binomial expansions, and trigonometric identities and series, as well as simple algebraic manipulations. These discussions are followed by sections on ways to validate a model by testing limits and by applying simple statistics. Last, but certainly not least, we include a reminder to check the reasonableness of any "answers" obtained from a model.

TAYLOR SERIES

Many numerical techniques are derived directly from the Taylor series; therefore, the Taylor series forms the foundation for approximation by polynomials of functions both known and unknown. Any function that is continuous and has derivatives may, in general, be expanded into a Taylor series. If we want to approximate a function $f(x)$ for values of x in a region near $x = a$, we write

$$\begin{aligned} f(x) \cong f(a) + (x-a)f'(a) + \frac{1}{2!}(x-a)^2 f''(a) \\ + \frac{1}{3!}(x-a)^3 f'''(a) + \cdots + \frac{1}{n!}(x-a)^n f^{(n)}(a) \end{aligned} \tag{4.1}$$

where $f'(a)$ indicates the first derivative, $f''(a)$ the second derivative, and $f^{(n)}(a)$ the nth derivative of $f(x)$ evaluated at the point $x = a$. The series (4.1) is called the Taylor series expansion of $f(x)$ in the neighborhood of $x = a$. The point $x = a$ should be chosen so that all derivatives of $f(x)$ at $x = a$ exist and are finite. In addition, if the difference $(x - a)$ is very small, only a few terms of the series need be used to effect a good approximation.

Suppose we want to evaluate $f(x)$ at some point, say $x = b$. We write the Taylor series expansion of $f(b)$, using Eq. (4.1), as

$$f(b) = f(a) + (b - a)f'(a) + \frac{1}{2!}(b - a)^2 f''(a) + \frac{1}{3!}(b - a)^3 f'''(a) + \cdots \tag{4.2}$$

If only the first term in Eq. (4.2) is used, then the function approximates to the constant $f(a)$ as shown in Figure 4.1a. The first two terms of Eq. (4.2) give an equation of a straight line from $f(a)$ with a slope $f'(a)$. This yields a value for $f(b)$ closer to the true value than that obtained from only the first term (see Figure 4.1b). The inclusion of three terms introduces curvatures due to $f''(a)$, so we come even closer to the real value of $f(b)$, as may be seen in Figure 4.1c.

The accuracy of our approximation for $f(b)$ increases with the number of terms used. The remainder, or error, after n terms in the Taylor series (4.1) is

$$R_n = |f^{(n+1)}(x)|_{\max} \frac{(|x - a|)^{n+1}}{(n + 1)!} \tag{4.3}$$

where the subscript "max" denotes the maximum magnitude of the derivative in the interval from $x = a$ to $x = b$. If an infinite number of terms in the Taylor series were calculated, there would then be no error, for the series would converge to the exact value of $f(x)$. How do we decide how many terms to use before the error becomes acceptable (or negligible)? The answer depends on the conditions of the problem at hand. Generally we are interested in values of $f(x)$ in the neighborhood of $x = a$, so we can approximate $f(x)$ very closely by using only a very few terms. However, for complicated functions that vary rapidly in value over a small range of the independent variable, more terms are required for a reliable approximation.

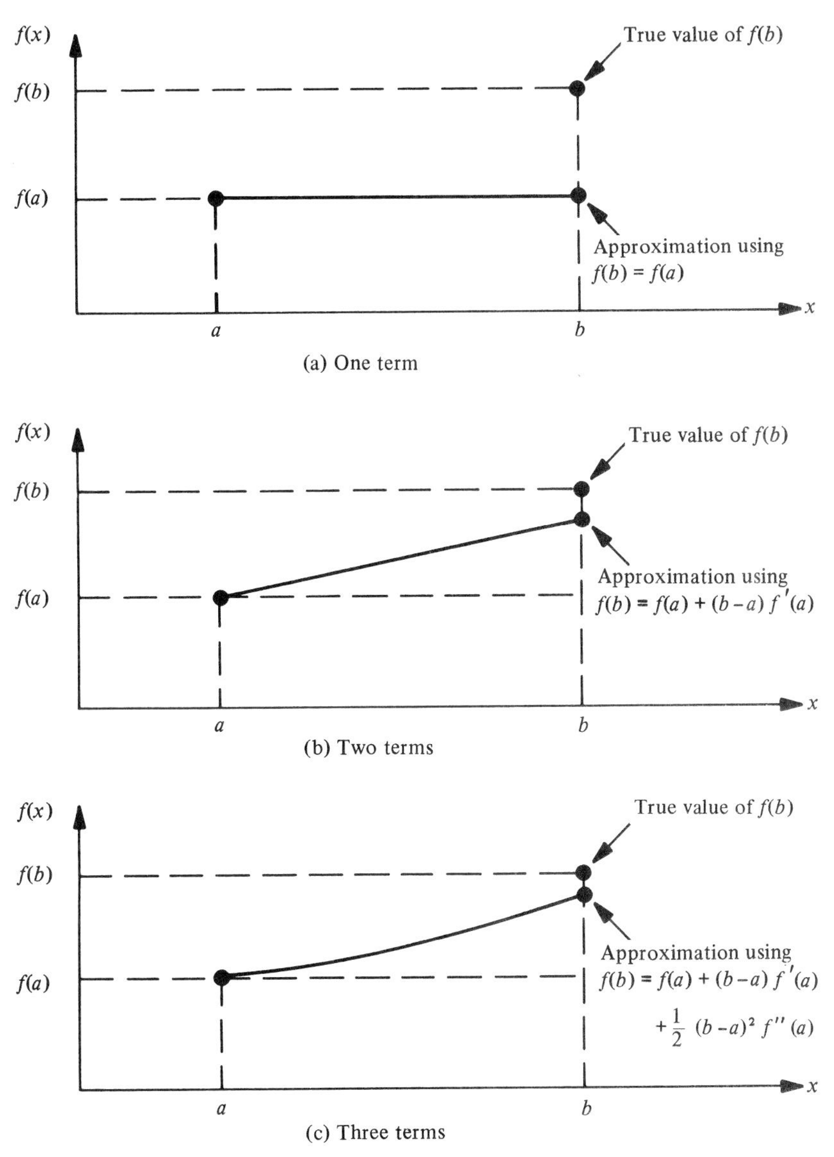

FIGURE 4.1
Approximations resulting from truncating Taylor series

BINOMIAL EXPANSION

Another example of a useful approximation technique is that of the binomial series expansion where

$$(a + x)^n = a^n + na^{n-1}x + \frac{n(n-1)}{2!}a^{n-2}x^2 + \frac{n(n-1)(n-2)}{3!}a^{n-3}x^3 + \cdots \tag{4.4}$$

This series converges for values of x such that $x^2 < a^2$; that is, we should think about using it if we have a binomial function of the form in Eq. (4.4) and $x^2 < a^2$. In Eq. (4.4) n is any number, positive or negative. If n is an integer, the series has a finite number of terms.

To illustrate the use of the binomial expansion we shall now look at two limiting expressions for the electric potential due to point charges. Any electric charge, or any group of electric charges, produces an electric field in the surrounding medium. The electric potential V at any point in such an electric field can be measured and can be shown to be directly dependent on the amount of charge setting up the field. The potential is also inversely proportional to the distance r from the measurement position to the charge (see Figure 4.2). For a point charge the

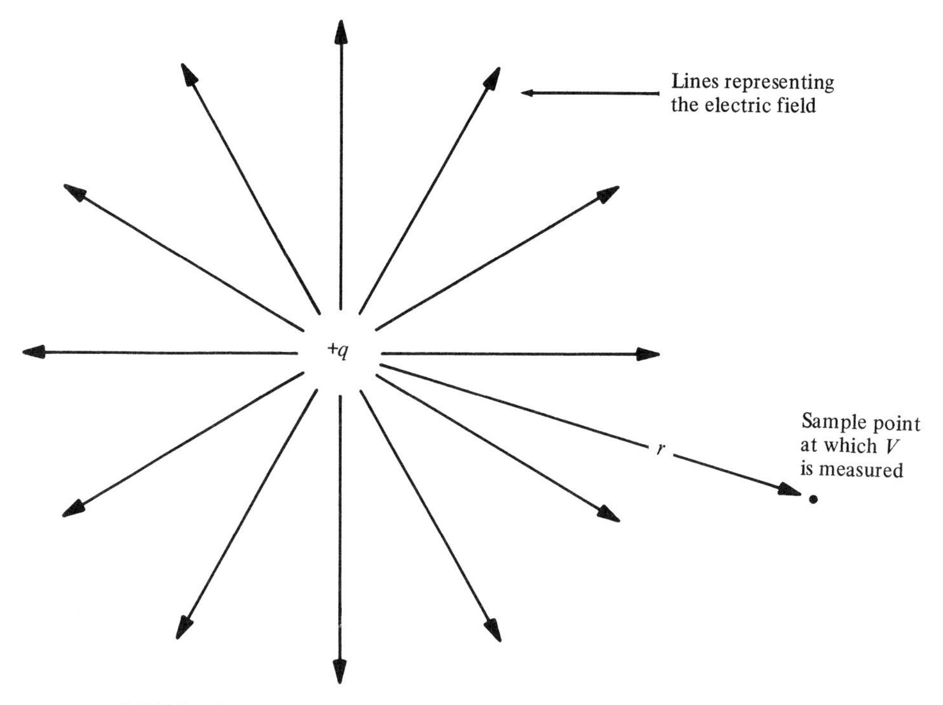

FIGURE 4.2
Electric field around a point charge

potential is

$$V = \frac{1}{4\pi\epsilon_0}\left(\frac{q}{r}\right) \tag{4.5}$$

where ϵ_0 is the permittivity constant.

Now we shift our attention somewhat and look at the electric potential for points on the axis of revolution of a uniformly charged disk. A circular disk of radius a is shown in Figure 4.3. Point P represents any point on an axis through the center of the disk. The equation for the electric potential V at any point P at distance r from the disk is

$$V = \frac{q}{2\pi a^2\epsilon_0}(\sqrt{a^2 + r^2} - r) \tag{4.6}$$

Here q represents the total charge on the disk, this charge being uniformly distributed over the area of the disk.

If Eq. (4.6) is to be valid for the special case of $r >> a$, it should reduce to Eq. (4.5) when $r >> a$. In other words, we would expect the disk to behave like a point charge when point P is very far away ($r >> a$) from the disk. Let us therefore approximate the radical in

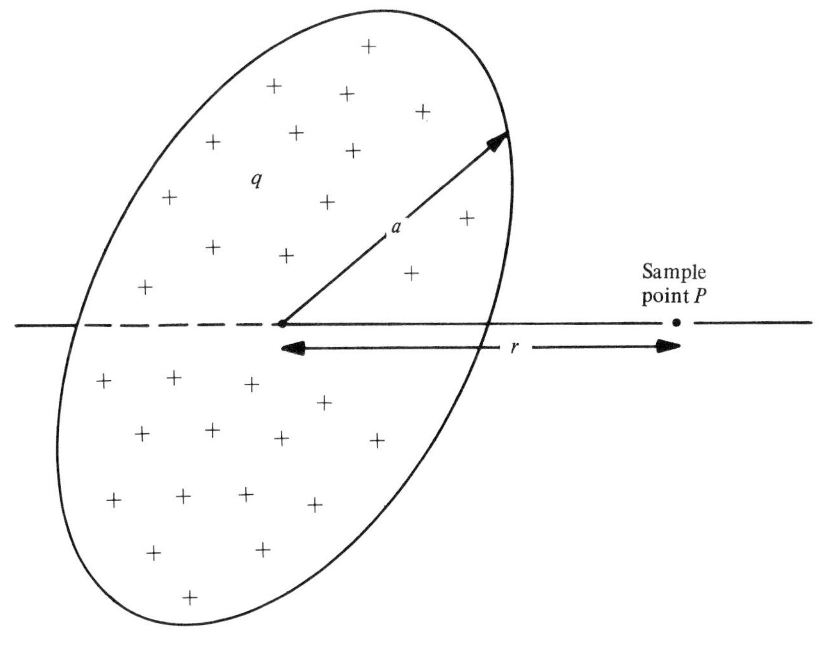

FIGURE 4.3
Uniformly charged circular disk of radius a

Eq. (4.6). We see that for $r >> a$

$$\sqrt{a^2 + r^2} \cong \sqrt{r^2} = r \tag{4.7}$$

which amounts to keeping only the first (or leading) term in a binomial series. But this result inserted into Eq. (4.6) yields a potential equal to zero when it should be the same as that for a point charge. Clearly, we have made an error! We need a better approximation of the quantity $\sqrt{a^2 + r^2}$ for large values of r. Let us rewrite the quantity $\sqrt{a^2 + r^2}$ in explicit binomial form, so that

$$\sqrt{a^2 + r^2} = r\left(1 + \frac{a^2}{r^2}\right)^{1/2} \tag{4.8}$$

We can then formally expand the quantity in parentheses by the binomial theorem and retain the next term in the series:

$$r\left(1 + \frac{a^2}{r^2}\right)^{1/2} = r\left(1 + \frac{1}{2}\frac{a^2}{r^2} + \cdots\right) \cong r + \frac{a^2}{2r} \tag{4.9}$$

Only two terms are retained because the third term involves a^4/r^4, and for $r >> a$ this term will be negligible. Equation (4.9) means that the potential V becomes

$$V \cong \frac{q}{2\pi a^2 \epsilon_0}\left(r + \frac{a^2}{2r} - r\right) = \frac{qa^2}{4\pi a^2 \epsilon_0 r} = \frac{1}{4\pi\epsilon_0}\frac{q}{r} \tag{4.10}$$

which is the result [see Eq. (4.5)] we expected.

It is worth underlining the reason that the first approximation [Eq. (4.7)] failed and led to an erroneous answer. We were using an approximation of $\sqrt{a^2 + r^2}$ to look at the *difference* between $\sqrt{a^2 + r^2}$ and r. When we are looking at small differences, as we are in fact doing in this limiting process, we must be careful in writing series to know what we are approximating and what is the *leading* term (the first nonzero term). In this problem we were actually trying to approximate not $\sqrt{a^2 + r^2}$ but the difference

$$\begin{aligned}\sqrt{a^2 + r^2} - r &= r + \frac{1}{2}\frac{a^2}{r} + \cdots - r \\ &= \frac{1}{2}\frac{a^2}{r} + \cdots\end{aligned}$$

So we must retain the term proportional to $1/r$ as the leading term in the expansion for the appropriate difference. We also note, incidentally, that such an approximation for a physical quantity far away from its "cause" is called a *far-field* approximation.

TRIGONOMETRIC SERIES

Trigonometric functions can also be expanded into series. Four of the more commonly used series are given below:

$$\sin x = x - \frac{x^3}{3!} + \frac{x^5}{5!} - \frac{x^7}{7!} + \cdots \tag{4.11a}$$

$$\cos x = 1 - \frac{x^2}{2!} + \frac{x^4}{4!} - \frac{x^6}{6!} + \cdots \tag{4.11b}$$

$$\sinh x = x + \frac{x^3}{3!} + \frac{x^5}{5!} + \frac{x^7}{7!} + \cdots \tag{4.11c}$$

$$\cosh x = 1 + \frac{x^2}{2!} + \frac{x^4}{4!} + \frac{x^6}{6!} + \cdots \tag{4.11d}$$

It is important to note here that the angle represented by x in the Eqs. (4.11) must be expressed in units of radians to ensure dimensional homogeneity.

We will now use a trigonometric series expansion to find an approximate expression for the sag of a tightly stretched string. Such a string, suspended between two fixed points a distance l apart, is shown in Figure 4.4. The elevation of both suspension points is the same. The string is shown hanging loosely in order to show the parameters representing the sag h, the height of the support y_1 above the horizontal axis, and the "catenary parameter" c. The equation of the catenary (that is, the form the string takes) with the vertex at $x = 0$ and $y = c$ is

$$y = c \cosh \frac{x}{c} \tag{4.12}$$

Referring to Figure 4.4 we can write the sag h as

$$h = y_1 - c \tag{4.13}$$

where

$$y_1 = c \cosh \frac{l}{2c} \tag{4.14}$$

since $x = l/2$ at the point of support. We now expand the hyperbolic

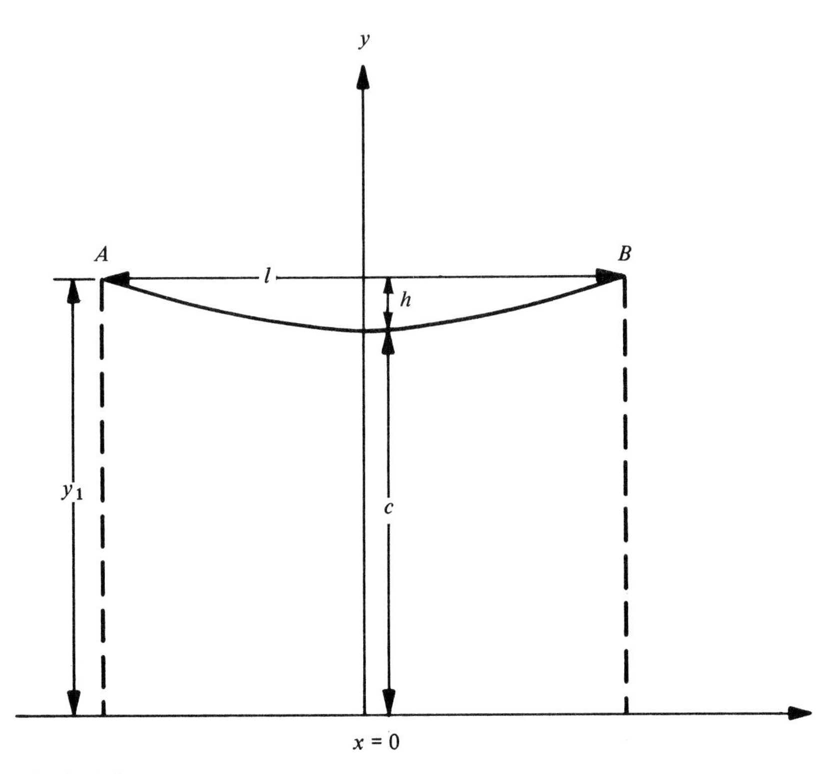

FIGURE 4.4
A string stretched between two fixed points at A and B

cosine in Eq. (4.14) using a trigonometric series. Then

$$h = c \cosh \frac{l}{2c} - c = c\left(1 + \frac{1}{2!}\frac{l^2}{4c^2} + \frac{1}{4!}\frac{l^4}{16c^4} + \cdots - 1\right) \qquad (4.15)$$

When the string is tightly stretched, h becomes very small and c is very large. Thus, we can neglect powers of l/c higher than the second (noting again that we are interested in a small difference), leaving the sag as

$$h = \frac{l^2}{8c} \qquad (4.16)$$

There is yet another approach to approximating with trigonometric series. Suppose we want to replace $\sin x$ with x to simplify an equation, as we shall do with the pendulum in Chapter 5. We then need to look at the limits on x for which $\sin x \cong x$ is a good approximation. With some quick calculations or with a quick look at some trigonometric tables (see also the problems at the end of the chapter), we see that for $x \leq \pi/6$ rad (30°), replacing $\sin x$ by x introduces a maximum error of 5%

[2% error for $x \leq \pi/12$ rad (15°)]. Thus, by limiting the range of x for which a relationship is valid, we are often able to simplify an equation by replacing a trigonometric function with an approximation. Similarly, the approximations of $\cos x \cong 1$ and $\tan x \cong x$ can be used with appropriate limits set on x.

ALGEBRAIC APPROXIMATIONS AND SIGNIFICANT FIGURES

When a term in an equation becomes significantly smaller than the other terms, we can often drop that term without introducing a large error. In Chapter 3 we discussed coefficients of thermal expansion in relation to the scaling of tightly fitting parts. Now we will look at the equations for two thermal expansion coefficients in greater detail to illustrate the nature of selectively dropping terms.

When temperature rises, the average distance between atoms in a heated body increases. This expansion occurs in any linear dimension of a solid, such as its length, width, and thickness. For a single dimension of length L_0, the increase in length ΔL is directly proportional to the rise in temperature ΔT. Hence, we can write

$$\Delta L = \alpha L_0 \, \Delta T \tag{4.17}$$

where α is the coefficient of linear expansion; it has different values for different materials. Rewriting Eq. (4.17) and replacing ΔL by $L - L_0$ allows us to write a relation for the heated length L:

$$L = L_0(1 + \alpha \, \Delta T) \tag{4.18}$$

Heating a plate or sheet of material causes both the length and the width to increase. Consider a rectangular plate whose length and width initially are L_0 and W_0, respectively. When heated, these dimensions become

$$L = L_0(1 + \alpha \Delta T)$$

and

$$W = W_0(1 + \alpha \Delta T)$$

Originally, the area of the plate was

$$A_0 = L_0 W_0$$

After heating the area becomes

$$A = LW = L_0 W_0 (1 + \alpha \Delta T)^2$$
$$= A_0[1 + 2\alpha \Delta T + (\alpha \Delta T)^2] \tag{4.19}$$

Coefficients of linear expansion are very small numbers (see Table 4.1). Therefore, the term α^2 in the brackets in Eq. (4.19) is extremely small compared to unity, so the term $(\alpha \Delta T)^2$ may be dropped. Hence

$$A \cong A_0(1 + 2\alpha \Delta T) \tag{4.20}$$

and we can define a surface coefficient of expansion γ as

$$\gamma \cong 2\alpha \tag{4.21}$$

where γ is thus derived from approximating $(1 + \alpha \Delta T)^2$ as $(1 + 2\alpha \Delta T)$.

Now we shift our attention to approximations in measurements, which brings us to the subject of significant figures. All measurements involve approximations and, therefore, they contain errors. We try to minimize these errors by reading and recording only those digits that can be read directly from a graduated scale, plus one estimated digit. Those recorded numbers are the *significant figures*. For example, the needle on the voltmeter in Figure 4.5 points to a value between 12 and 13 V. The first number to record is thus 12. Since there are no graduations on our scale that divide the interval between 12 and 13 V into equal segments, we must estimate the third digit. If we see the needle as 0.2 of the way from 12 to 13, we can then properly record our voltage measurement as 12.2 V. This gives us a reading "good" to three significant figures because it contains two digits read directly from the graduated scale without question plus one that is estimated.

An important fact to remember about significant figures is that it is *not* the decimal point that determines the number of significant figures.

Table 4.1
Coefficients of Linear Expansion

Substance	α [(°C)$^{-1}$]
Aluminum	24×10^{-6}
Brass	20×10^{-6}
Copper	14×10^{-6}
Glass	$4\text{–}9 \times 10^{-6}$
Steel	12×10^{-6}
Zinc	26×10^{-6}

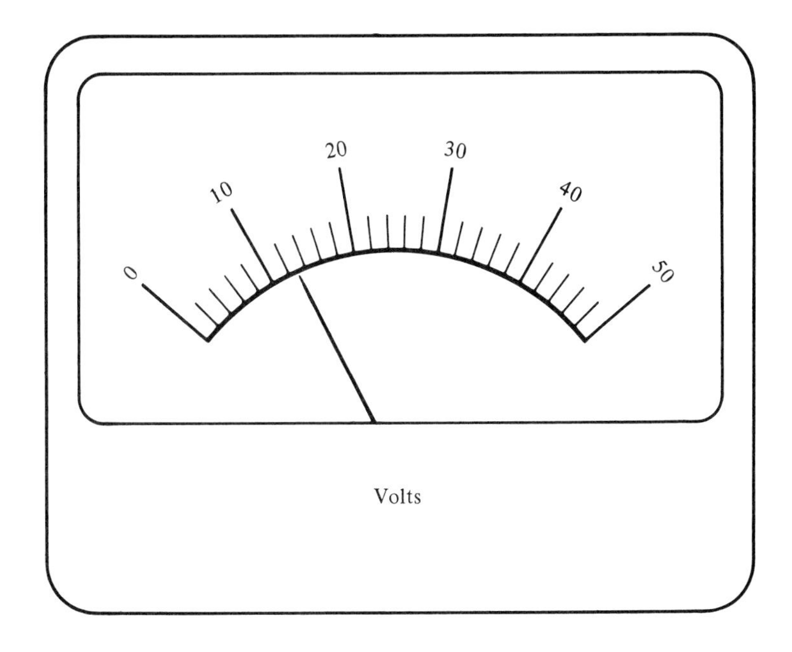

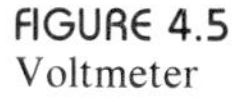
FIGURE 4.5
Voltmeter

Suppose our scale on the voltmeter shown in Figure 4.5 had been from 0 to 5 V instead of 50 V. Then our reading of the needle would have been 1.22 V, and our measurement would still have contained just three significant figures. Table 4.2 contains some examples of significant figures with comments to help clarify some of the confusion that sometimes accompanies the recording of significant figures.

There are additional rules and conventions that have been adopted with regard to the use of significant figures. When adding or subtracting a set of measurements, the number of columns to be added or subtracted beyond the decimal point is determined by the measurement having the smallest number of significant decimal places. This is illustrated by the following examples:

$$\begin{array}{r} 53.24 \\ +\ 3.333 \\ +\ 2.4 \\ \hline 58.9 \end{array} \qquad \begin{array}{r} 489.3213 \\ -\ \ \ 5.487 \\ \hline 483.834 \end{array}$$

When multiplying or dividing, the product or quotient cannot have more significant figures than the lowest number of significant figures of

Table 4.2
Examples of Significant Figures

Measurement	Comments	Significant figures
63.2	Clear	Three
6.32	Clear	Three
0.00632	Clear	Three
6.32×10^5	Clear	Three
0.041	Clear	Two
0.0410	Confusing ⟶	Two (0.041) *or* three (0.0410)
0.00008	Clear	One
9415	Clear	Four
9400	Confusing ⟶	Two (94×10^2) *or* three (940×10^1) *or* four (9400)
52.0	Clear	Three

any other term in the calculation. In other words, the results of calculations using a set of measured values is only as accurate as the least accurate value employed. Consider the following examples:

$$21.982 \times 3.72 = 81.77304 \rightarrow 81.8$$
$$101.527 \times 0.0031 = 0.3147337 \rightarrow 0.31$$
$$789.30 \div 0.05 = 15786 \rightarrow 2 \times 10^4$$

Another important question is when (or if) to round off numbers. Generally, it is desirable to round off at the completion of calculations. Indeed, the advent of the hand calculator has made this normal procedure. Dropping nonsignificant digits before the final computation is reached usually increases the uncertainty in the answer. A convention that is widely used in round-off calculations uses the number 5 as a benchmark. Numbers less than a 5 following the last significant digit to be retained are dropped or replaced by a 0, while numbers greater than 5 result in adding 1 to the preceding number. If the first digit to be rounded is itself a 5, then the preceding digit is left as is if it is even, and it is raised to the next even digit if it is odd. Thus, for example,

$$5.017 \rightarrow 5.02$$
$$5.015 \rightarrow 5.02$$
$$5.014 \rightarrow 5.01$$
$$5.025 \rightarrow 5.02$$

A discussion of significant figures would not be complete without some mention of numbers with unlimited significant figures. Some whole numbers represent an exact count and, as such, contain an unlimited number of significant figures. They are usually written with no digits after the decimal point; they may or may not have the decimal point, depending on their use. To indicate an exact number, we may write 35. or, in the case of the formula for the circumference of a circle,

$$C = 2\pi r$$

the number 2 represents an exact count and does not have a decimal point. Other examples of terms with unlimited significant figures are π and e, the base of Napierian logarithms. However, if we write 35.0 rather than 35., then we have three significant figures, indicating we have measured to the first decimal place.

The use of significant figures so as to minimize uncertainty requires experience and judgment. We must keep in mind the significance of the data from which we derive our results and try to determine their validity in light of the assumptions made.

VALIDATING THE MODEL—ERRORS

When building mathematical models we constantly deal with numbers, numbers that often come from experimental observations, and experimental observations are always somewhat inaccurate or in error. We discussed one source of error in the previous section on significant figures. Reading a dial or scale on an instrument requires an estimate of the last digit recorded. Such an estimate introduces inaccuracy into the measurement. If several observations are used in calculating a result, we must know how the inaccuracies of the individual measurements contribute to the inaccuracy of the result. Therefore, we will devote the rest of this chapter to a brief discussion of some simple statistics.

Error is the difference between a measured value and the true or exact value. How much error is introduced depends on the sensitivity and accuracy of the measuring device as well as on the ability and skill of the observer. No matter how skillful we are as observers, no matter how precise our instrumentation may be, error is *always* present. For this reason, every analysis of experimental results should account for the errors involved.

There are two basic types of errors, *systematic* and *random*. When an observed value deviates from the true value in a consistent way, we have *systematic errors*. A meter, such as an ammeter, that is not properly zeroed introduces systematic error. If the needle on an indicating scale is bent, observed values will deviate from the true value by a consistent amount. Less obvious systematic errors occur when equipment is not properly calibrated. Instruments do vary their output with use; they therefore require careful calibration before a set of measurements is taken. If this calibration is done, and if necessary corrections are applied to subsequent readings, systematic errors can be reduced (but never entirely eliminated).

Random errors are produced when repeated observations of a quantity are made. They arise because most experimental situations have a large number of unpredictable and unknown sources of inaccuracy. For example, a thermocouple junction may become corroded or loose; the ambient temperature in a room or the power line voltage can change; or dirt may cause unexpected friction between moving parts. The resulting random error varies in magnitude, with both positive and negative values occurring in a random sequence. The distribution of such random errors follows statistical laws.

A useful statistic is the *percentage error,** which can be defined as

$$\%\ \text{error} = 100 \cdot \left[\frac{\text{expected result} - \text{measured value}}{\text{measured value}}\right]$$

or

$$\%\ \text{error} = \frac{100 \cdot (X_e - X_m)}{X_m} \tag{4.22}$$

where X_e is the expected result and X_m is the measured value. Here we use the term "expected result" to mean the true value when we know what that is or the calculated value when we can predict our results using equations. Consider a systematic error of +2 amperes (A) in all readings of an ammeter with a bent needle. When the meter reads 100 A we have a percentage error of [using Eq. (4.22)]

$$\%\ \text{error} = \frac{100 \cdot (102 - 100)}{100} = 2\%$$

However, when this same ammeter reads 20 A, the percentage error becomes

$$\%\ \text{error} = \frac{100 \cdot (22 - 20)}{20} = 10\%$$

Here, then, is another example of how scaling affects results!

*We use implicitly the definitions of *absolute error* $(X_e - X_m)$ and *relative error* $((X_e - X_m)/X_m)$ in obtaining the percentage error.

Earlier in this chapter we discussed the percentage errors introduced when truncating a series. For example, when $\theta = \pi/12$ rad (15°), the percentage error incurred by allowing $\sin x \cong x$ is, using Eq. (4.22),

$$\% \text{ error} = 100 \cdot \left[\frac{\sin \pi/12 - \pi/12}{\pi/12}\right] = -1.138\%$$

We note here that *errors* and *mistakes* are not the same thing. An *error* has already been defined as the difference between a measured value and the true value. We have also noted that some errors are unavoidable. *Mistakes*, on the other hand, are blunders made by the person performing the experiment. Blunders may be such mistakes as incorrectly reading instruments, erroneously recording numbers, making arithmetic errors in calculating results, or taking data before the instruments are thoroughly "warmed up." These types of inaccuracies *can* be avoided by working carefully and meticulously.

Given that we will have to deal with systematic and random errors as well as an occasional blunder, can we estimate how these inaccuracies affect the accuracy and precision of our measurements? We shall define *accuracy* as a representation of how exact our measured value is when compared either to the true value or to a known input, that is, how near we are to the established value. Accuracy is usually expressed as a percentage of a full-scale reading. If we are reading from a 100-V scale voltmeter with an accuracy of 5%, our values would be accurate to within ±5 V. *Precision* is the ability to reproduce a set of measurements with a given accuracy. The more precise a set of readings, the nearer the individual readings are *to each other.* To see the distinction between precision and accuracy, consider the measurement of a (known) voltage of 50 V with the meter just mentioned. Our readings are 54, 53, 55, 53, and 55 V, all of which fall within our accuracy range of ±5 V. Here the precision is ±1% because the maximum deviation from the mean reading of 54 V is only 1 V. Our data, then, are very reproducible but not very accurate. Figure 4.6 illustrates these differences.

In addition, the *sensitivity* of a measuring device controls its accuracy because the sensitivity identifies the minimum amount of change that the device is capable of detecting and indicating. Suppose we want to measure very small voltages, say less than 1 millivolt (mV). Our voltmeter allows us to use any one of three ranges: 0–50 V, 0–2.5 V, or 0–5 mV. If we use the 2.5-V range (scale), the movement of the needle with a 0.5-mV change in voltage cannot be discerned. However, with the range set from 0 to 5 mV, we will see a noticeable, and readable,

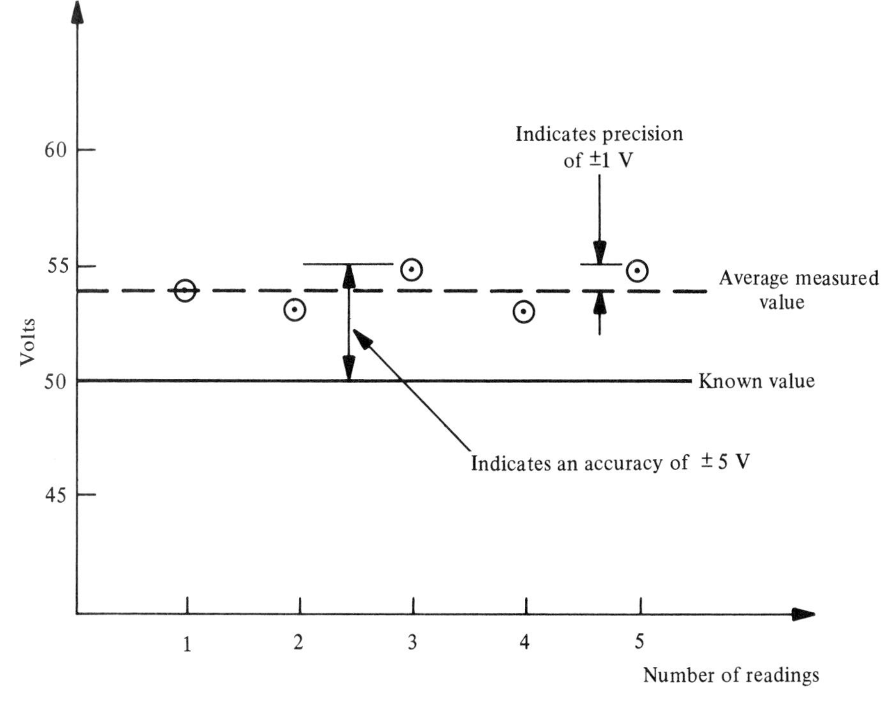

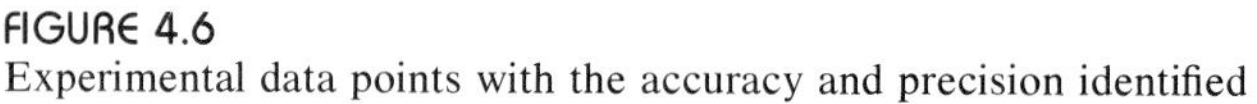

FIGURE 4.6
Experimental data points with the accuracy and precision identified

change in the position of the needle on the scale. Changing from the 0–2.5 V scale to the 0–5 mV scale has made our voltmeter much more sensitive, and it therefore allows us to record more accurate readings. Here scaling controls sensitivity.

Figure 4.7 illustrates the relationships between errors, blunders, experimental apparatus, and observers. Often the distinction between accuracy and precision is related to the differences between systematic and random errors. High accuracy usually indicates a measurement with small systematic errors, whereas high precision usually represents a measurement with small random errors.

AVERAGING—MEAN, MEDIAN, MODE

We often want to approximate a value by using an average. There are several ways to look at the concept of "average." The one we are most

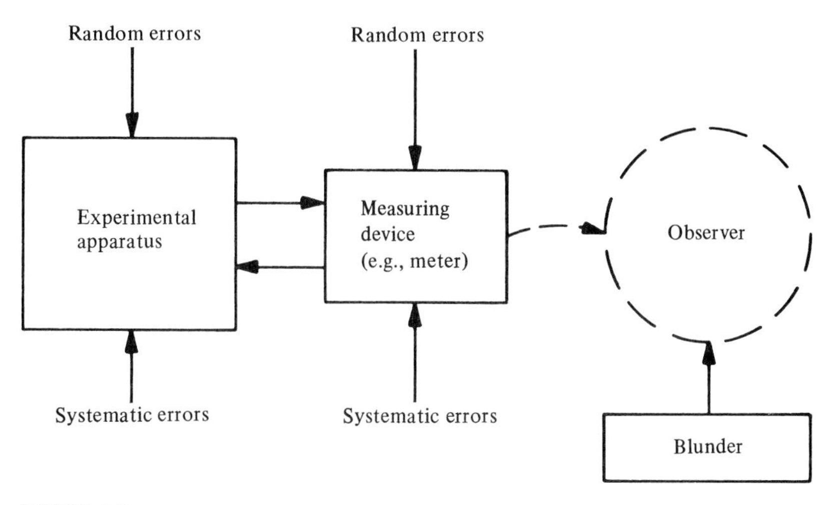

FIGURE 4.7
Schematic diagram of the introduction of error into measurements

familiar with is the arithmetic average or *mean*. Two other common averages are the *median* and the *mode*.

As shown in Figure 4.6, individual measurement readings (data points) usually vary from each other. The observer wants to find the best estimate of the magnitude of the readings. We will define such a "best estimate" as the mean or average of n measurements. The *mean* of a set of numbers is simply the sum of all the numbers x_i divided by the number of them, or in mathematical notation,

$$\bar{x} = \frac{1}{n}\sum_{i=1}^{n} x_i = \frac{x_1 + x_2 + x_3 + \cdots x_n}{n} \tag{4.23}$$

where the bar over the variable x indicates the mean value of x, and the symbol

$$\sum_{i=1}^{n}$$

is read "the sum from $i = 1$ to $i = n$" readings.

Suppose we have taken 100 measurements of the (outdoor) traffic noise levels near a school. Several children walk by the microphone during the time we are measuring and each of them makes a very loud noise near the microphone. The average sound pressure level reported is the level exceeded 50% of the time, or for 50 of our readings, since they are taken at equal time intervals. This is an example of a *median* value, denoted here by L50. It removes the bias introduced into a mean value by the few very loud noises. In other words, the median value is that measured value exceeded by 50% of our measurements. To find that median value, we put all of our measured values in order of decreasing (or increasing) magnitude, and then we count off from the top that number of readings equal to $n/2$. The value of the "$(n/2)$th" reading is the median value of that set of measurements.

The *modal* value in any set of observations is the one value that is observed most often. Table 4.3 lists a set of 100 noise level readings taken at 15-sec intervals and indicates the values for the mean, median, and mode. (Note that the units for noise levels are decibels or dB's.)

Having taken a set of measurements, plotted them, and calculated the mean or "best value," we would now like to find a quantitative way to express the dispersion or spread of the individual measurements about the mean. A useful indicator of scatter about the mean is the *standard deviation*. This is the accepted statistical measure of dispersion. To find the standard deviation, usually symbolized by σ, we compute the difference between any measurement x_i and the mean $\bar{x}$ of the set. Since some of these differences are positive and some are negative, we square all of these values of $(x_i - \bar{x})$, and then we sum these squared values. We thus have only positive numbers to add.

Table 4.3
Noise Level Measurements in Decibels for $n = 100$

	Decibels	Number of readings
	90–91	× × × × × × × ×
	88–89	× × × × ×
	86–87	× × ×
	84–85	× ×
	82–83	× × × ×
	80–81	× ×
	78–79	× × ×
	76–77	
	74–75	× × × × × ×
	72–73	× × × ×
Mean	70–71	× × × × × ×
	68–69	× × ×
Median	66–67	× × × × × × ×
	64–65	× × × × ×
	62–63	× × × × × × × × ×
Mode	60–61	× × × × × × × × × × × × × ×
	58–59	× × × × × ×
	56–57	× × × × × × × ×
	54–55	× × × ×
	52–53	×
	50–51	

Next, we take the average of the squares by dividing by n, and then we take the square root of this result. We thus have

$$\sigma = \left[\frac{1}{n}\sum_{i=1}^{n}(x_i - \bar{x})^2\right]^{1/2} \tag{4.24}$$

Clearly, σ is small if the x_i deviate very little from the mean. A small standard deviation generally indicates high precision in our measurements. How does this rleate to the accuracy of the mean $\bar{x}$? In many cases the error in $\bar{x}$ is not likely to be greater than $\sigma/n^{1/2}$. Thus, the more measurements we have, the more reliable the mean becomes.

CURVE FITTING: LEAST SQUARES METHOD; THE CONTINUUM HYPOTHESIS

We pointed out earlier (Chapter 3) that the most convenient form for the presentation of the results of an experiment is often a graph. When plotting a graph, we normally would not draw a curve through all of the data points since there are uncertainties in the measurements recorded. Thus, we need to decide what method to use to draw a curve that will be a "best fit" for the data points. In other words, we are approximating again!

If the accuracy of the curve is not too important, and if we only want a qualitative idea of how one variable depends on another, then we can draw the curve by eye. When doing this we are (usually) assuming that one variable is a continuous function of the other, so we draw a smooth curve. It is sometimes helpful to "distribute" the number of points above and below the line equally, as shown in Figure 4.8.

When we need to *interpolate* (estimate between measured values) or *extrapolate* (estimate beyond the range of measured values), greater accuracy in curve fitting is necessary. In the case of extrapolation, errors in drawing the curve are magnified in the final result, so that it is important to obtain an accurate curve. Extrapolation is most accurate when the curve is a straight line.

The most commonly used method of obtaining the best straight line through a series of points is the *method of least squares*. We assume in using this method that all the scatter (variation of the data from the drawn curve) in the points derives from errors in measuring only one of the variables. It is customary to plot the variable introducing the most scatter as the ordinate. Then the best straight line is the one that minimizes the errors in the vertical direction.

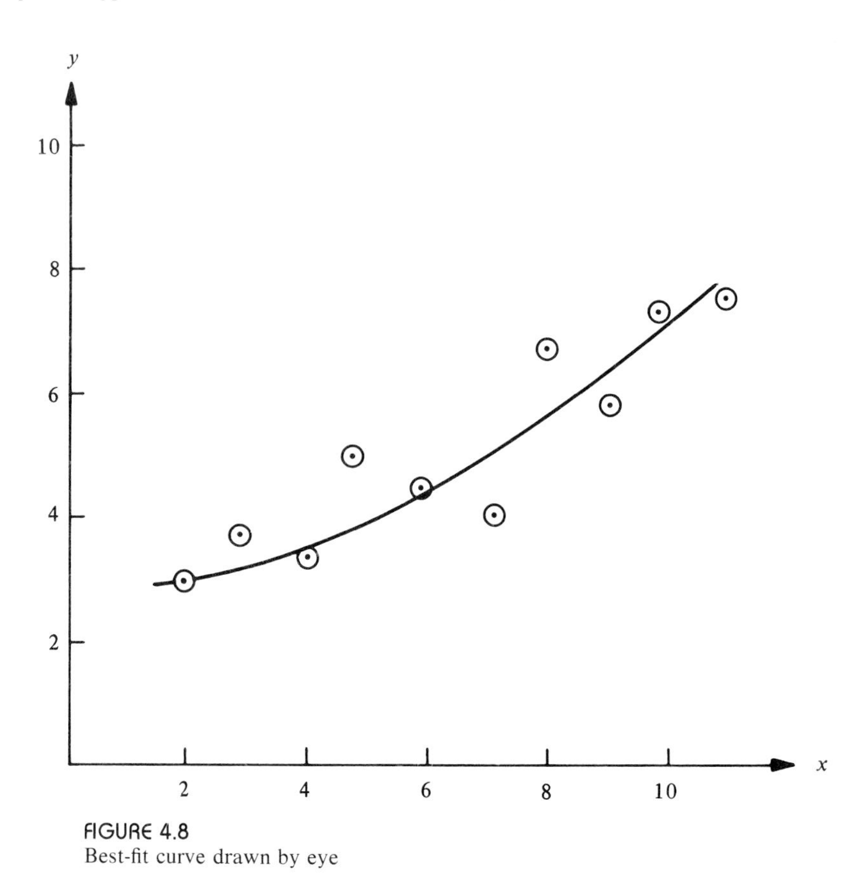

FIGURE 4.8
Best-fit curve drawn by eye

We are thus looking for an equation of the form

$$y = mx + b \tag{4.25}$$

where m is the slope of the straight line and b is the y intercept. To minimize the error in the ordinate at all points x_i, we first must sum the squares of the distance errors E_{y_i} in the y direction:

$$S = \sum_{i=1}^{n} E_{y_i}^2 = \sum_{i=1}^{n} [y_i - (mx_i + b)]^2 \tag{4.26}$$

Next we differentiate the right-hand side of Eq. (4.26) with respect to both m and b, the unknown constants, and equate the derivatives to zero. These operations result in the following:

$$\frac{\partial S}{\partial m} = \sum_{i=1}^{n} [2(y_i - mx_i - b)(-x_i)] = 0$$

$$\sum_{i=1}^{n} x_i(y_i - mx_i - b) = 0 \tag{4.27}$$

and

$$\frac{\partial S}{\partial b} = \sum_{i=1}^{n} [2(y_i - mx_i - b)(-1)] = 0$$

$$\sum_{i=1}^{n} (y_i - mx_i - b) = 0 \tag{4.28}$$

Solving Eqs. (4.27) and (4.28) simultaneously yields the sought-after values of slope m and intercept b:

$$m = \frac{n\sum_{i=1}^{n} x_i y_i - \left(\sum_{i=1}^{n} x_i\right)\left(\sum_{i=1}^{n} y_i\right)}{n\sum_{i=1}^{n} x_i^2 - \left(\sum_{i=1}^{n} x_i\right)^2} \tag{4.29}$$

$$b = \frac{\left(\sum_{i=1}^{n} y_i\right)\left(\sum_{i=1}^{n} x_i^2\right) - \left(\sum_{i=1}^{n} x_i y_i\right)\left(\sum_{i=1}^{n} x_i\right)}{n\sum_{i=1}^{n} x_i^2 - \left(\sum_{i=1}^{n} x_i\right)^2} \tag{4.30}$$

We will now do an example using the method of least squares. Table 4.4 lists data obtained in an experiment. We want to plot y as a linear function of x, where our resulting line will be the best fit. Thus, we need to find m and b in Eq. (4.25). First we calculate the terms

Table 4.4
Experimental Data Points

y_i	x_i
1.0	0
2.1	1.
2.8	2.
3.6	3.
5.0	4.
5.5	5.
8.0	6.
6.4	7.
$\Sigma y_i = 34.4$	$\Sigma x_i = 28$

$x_i y_i$	x_i^2
0	0
2.1	1.
5.6	4.
10.8	9.
20.0	16.
27.5	25.
48.0	36.
44.8	49.

and

$$\sum_{i=1}^{n} x_i y_i = 158.8 \qquad \sum_{i=1}^{n} x_i^2 = 140.$$

Then, using Eqs. (4.29) and (4.30), we find

$$m = \frac{8(158.8) - (28)(34.4)}{8(140) - (28)(28)} = \frac{307.2}{336} = 0.91$$

and

$$b = \frac{(34.4)(140) - (158.8)(28)}{8(140) - (28)(28)} = \frac{369.6}{336} = 1.1$$

Hence, the best straight line is given by

$$y = 0.91x + 1.1 \tag{4.31}$$

and is shown in Figure 4.9. The least squares technique can also be applied to obtain curves other than straight lines, for example, higher-order polynomials.

Experimental results may also be displayed graphically by using a *histogram* or *bar chart* where one coordinate indicates the values recorded and the other coordinate indicates the frequency of occurrence. The histogram shown in Figure 4.10 displays exactly the same data given in Table 4.3. The noise levels were recorded in 2-dB windows; that is, the number of readings registering 60 dB plus the number of readings registering 61 dB makes up the total recorded for that interval as 14. In Figures 4.11a and 4.11b we have plotted the same data (from Table 4.3) for 6-dB and 10-dB intervals, respectively. When plotting the observations in 10-dB intervals (Figure 4.11b) we remove any chance of distinguishing variations in noise levels that occur within that 10-dB interval. We thus "lose" the fact that several children shouting near the microphone caused a larger number of readings around 90–91 dB than either just above or just below that level. As the plotting

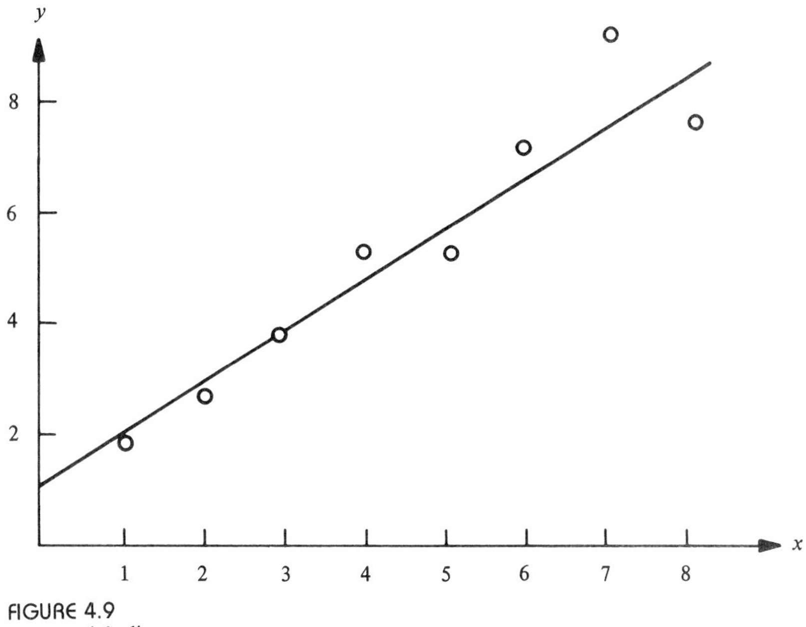

FIGURE 4.9
Best straight line

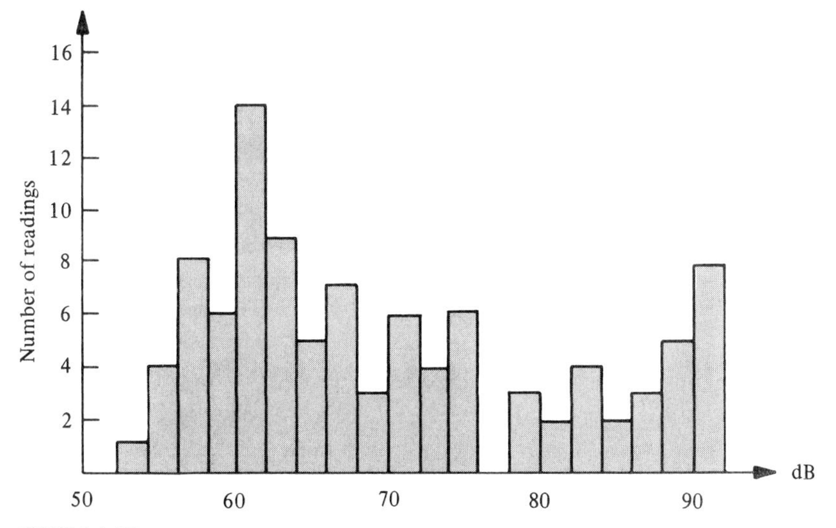

FIGURE 4.10
Histogram of noise level measurements

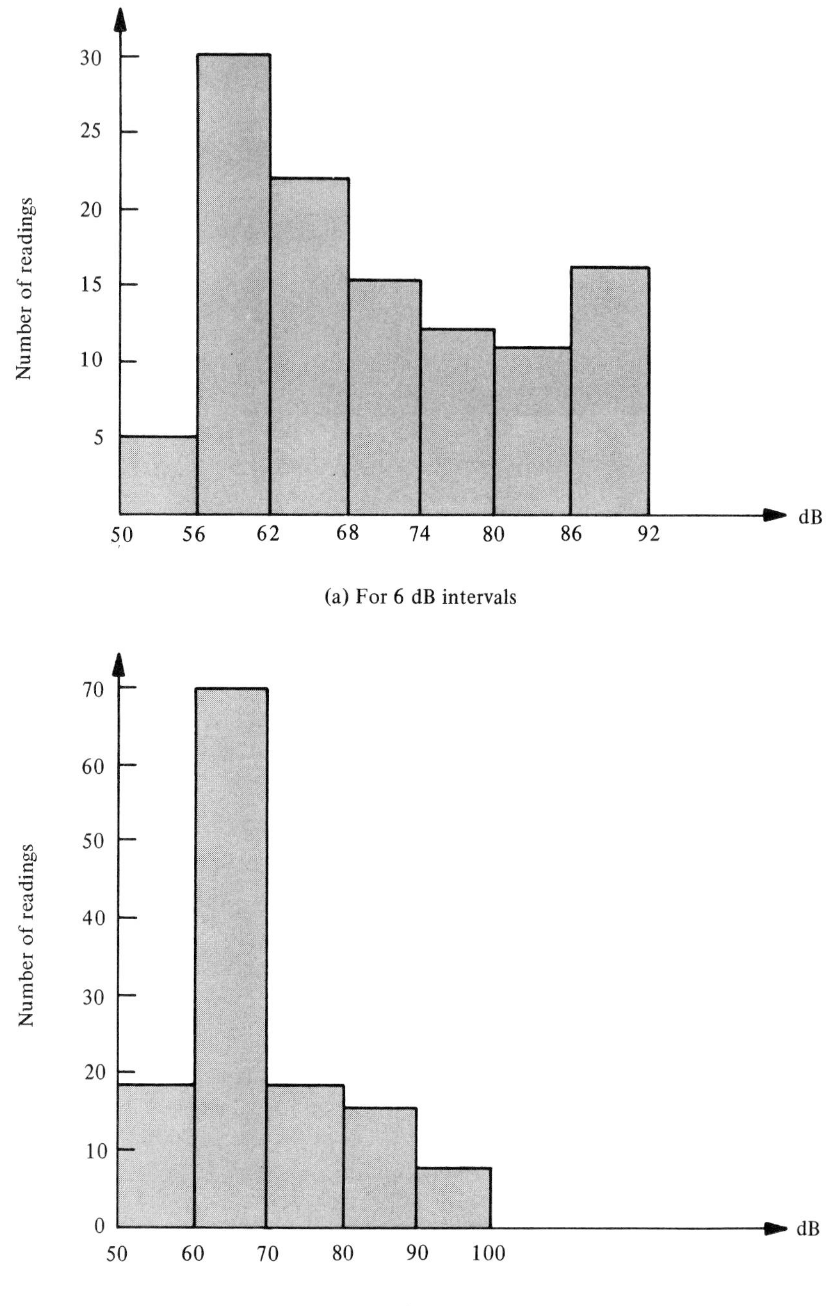

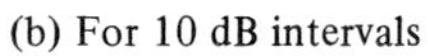
(b) For 10 dB intervals

FIGURE 4.11

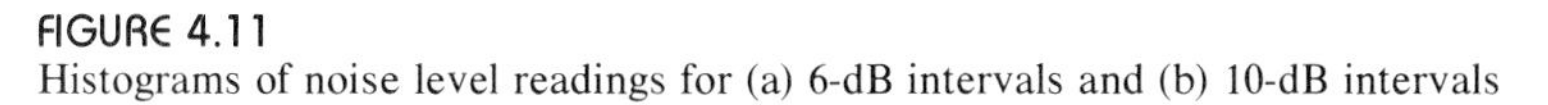
Histograms of noise level readings for (a) 6-dB intervals and (b) 10-dB intervals

interval is shortened to 6 dB, fluctuations occur. With the 2-dB interval (Figure 4.10) we see even more fluctuations in the histogram. If we can find a measuring interval that is

1. *large enough* so that sufficient readings fall in that interval (to eliminate excess fluctuations), and
2. *small enough* so that occasional loud or soft noise levels provide some variations (to avoid excess smoothing by averaging over too long an interval),

then we can *approximate* the steplike histogram in Figure 4.11a by a continuous function, as illustrated in Figure 4.12. The choice of a measuring interval that fits the foregoing criteria can best be illustrated by example. Consider again the data presented in Table 4.3, where we

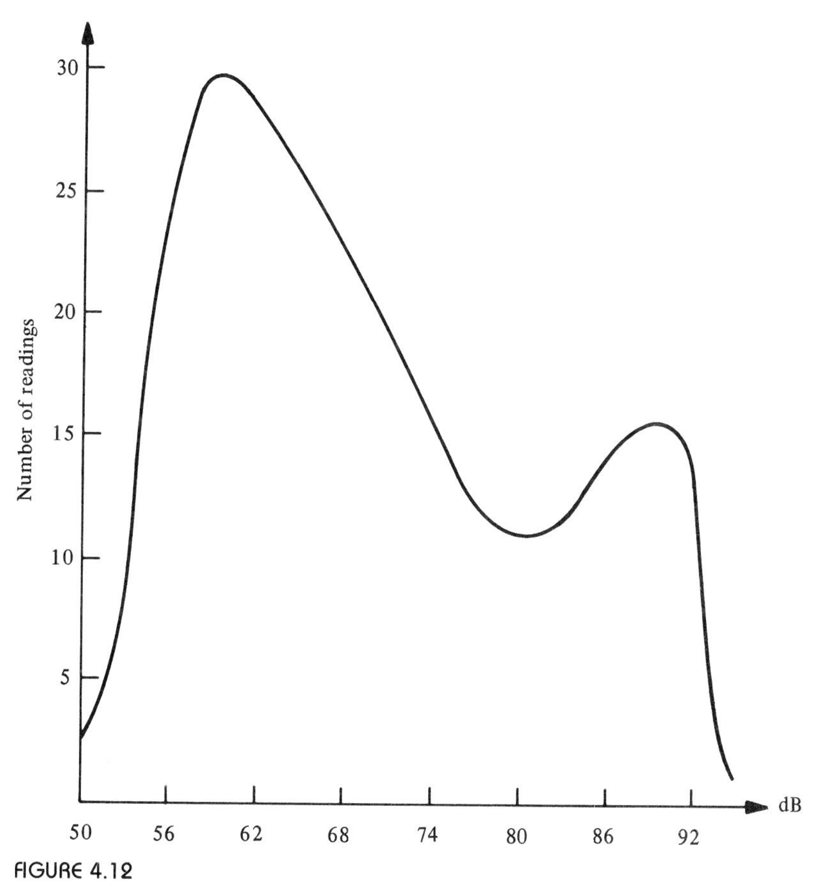

FIGURE 4.12
Continuous plot of noise level as a function of the number of readings

have a decibel range of 52–91 dB subdivided into 20 equal intervals of 2-dB size.

If we wish to *approximate* the number of readings as a continuous function of noise level, the number of readings must be measured over intervals that are neither too small nor too large. (The assumption that such discrete data can be plotted as a continuous curve is often referred to as the *continuum hypothesis*.) The measuring interval must be large enough so that a significant number of readings are included, but small enough that variations in the readings themselves within each interval will show up. To aid us in our choice of appropriate measuring interval, we will make a chart (Table 4.5) of the data in Table 4.3 from which we can plot the density of readings against the length of measuring interval. Table 4.5 uses an interval of m dB centered around a level of 66 dB.

Figure 4.13 displays a plot of the density of the readings as a function of the length of the measuring interval. Note that if the length of the measuring interval is less than 4 dB, the density fluctuates wildly and is not representative of the complete picture. Similarly, if the measuring interval is greater than 8 dB, the density curve is smoothed out so that real fluctuations are averaged out. Thus, a measuring interval in the range of 4–8 dB would be appropriate for approximating the number of readings as a continuous function of the noise level (see Figure 4.11a).

No definite criteria have been offered here as aids in choosing a suitable measuring interval. Indeed, there are none. Thus, when we wish to approximate discrete data with continuous functions, we must draw a curve similar to Figure 4.13, and then we look for those measuring intervals where the density of data points neither fluctuates too wildly nor smooths out too much. Making such choices becomes easier

Table 4.5
Chart Developed from Data in Table 4.3 for Use in Determining Appropriate Measuring Interval for Approximating with a Continuous Curve

Length of interval, m dB	1	2	3	4	5	6	7	8	10	20	30	40	50
Interval, $66 \pm m/2$ dB	66.5 65.5	67 65	67.5 64.5	68 64	68.5 63.5	69 63	69.5 62.5	70 62	71 61	76 56	81 51	86 46	91 41
Number of readings in the interval[a]	6	9	12	13	17	19	24	27	37	68	78	95	100
Density[b]	6	4.5	4	3.25	3.4	3.17	3.43	3.34	3.7	3.4	2.6	2.38	2.00

[a]Taken from Table 4.3.
[b]Average number of readings per 1-dB window in the interval.

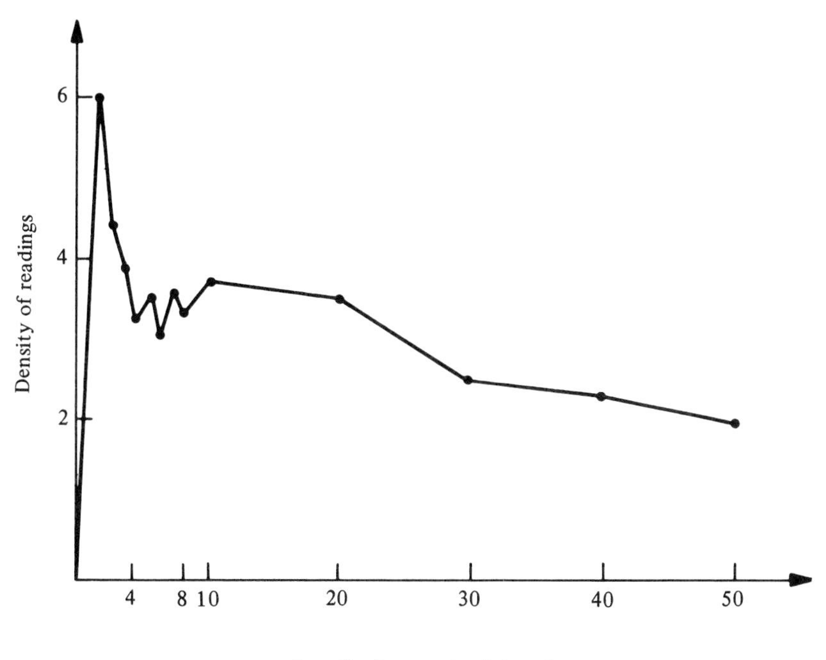

FIGURE 4.13
Variation in the density of readings' dependence on measuring interval

with practice. We will use this continuum hypothesis again in Chapter 7 when we discuss car-following models.

It is also possible to draw smooth curves and histograms by computer. When we need to generate many graphs from raw data, it is often more convenient to have a computer do it for us. Computers can be programmed to record raw data, process it, and display the results graphically. In such a process both errors and approximations are built into a program. Great care must thus be exercised in the selection of a program, whether it is packaged or written especially for a particular investigation.

VALIDATING THE MODEL—TESTING

There are several ways to validate a mathematical model through testing. We met one method earlier in this chapter when we discussed the electric potential at a point in an electric field, the field being set up by a uniformly charged disk. That was the method of testing an equation by evaluating it in the limit. If a model reduces to a known function that is representative of the model in the limit, then that model has passed one validation test. In our earlier discussion of electric potential, we validated Eq. (4.6) by showing that it reduced to Eq. (4.5) for a point charge when the potential was being measured at a point far enough away from the disk that it would appear to act as a point charge.

A mathematical model may also be tested by taking experimental data. Suppose we derive a model to predict how sound will travel over a certain ground surface under different weather conditions. A way to validate our model would be to devise and carry out an experiment to measure sound propagation over the ground surface in question and under specific weather conditions. Then we can compare our data with values predicted by our mathematical model. How close we may come to validating our model depends on how good our model is as well as on our experimental technique. We have already discussed some factors affecting experimental measurements earlier in this chapter. Figures 4.8 and 4.9 could also represent curves drawn from a model, with the validating experimental values scattered about them.

If we know roughly what the magnitude of our results should be, then we can test our model to see if it provides us with reasonable values. This involves using known or (on occasion) experimentally determined variables in our mathematical model. When such values are "plugged in" to our model equations, the results give us some indication of the validity of our model. If the results are similar to what we know is the expected value, then we are usually headed in the right direction with our model development. If not, back to the drawing board!

SUMMARY

We have devoted this chapter to a discussion of approximation techniques in model building and of model validation procedures, including some statistical treatment of experimental data. We have shown how truncating a series expansion can provide us with an approximate expression such as that found for the sag of a tightly stretched string. We also derived the surface coefficient of thermal expansion by using an algebraic approximation. The importance of the proper use of significant figures was discussed in relation to minimizing errors in experimental measurements.

Working with mathematical models means that we are constantly dealing with numbers that come from experimental observations. Such numbers always involve inaccuracies, or errors. We discussed both random and systematic errors, as well as how they affect the precision and accuracy of any set of measurements. A brief look at how we could quantify inaccuracies introduced us to the concepts of mean, median, mode, and standard deviation. Curve fitting techniques were shown to be additional approximation techniques. Illustrative examples were worked out using the least squares method and the continuum hypothesis. Finally, we showed that our models can be validated by testing in the limit, by taking experimental data, and by checking for the reasonableness of our answers.

We ought to note, too, that much of the material presented in this chapter (particularly the last half) will not be used in later developments in this book. However, this material is logically relevant to the present discussion, so we have included it here for the sake of completeness, and in the hope that it will be useful to the reader in related endeavors.

Problems

4.1 The readings of a certain voltmeter are subject to systematic error because the needle is mounted in such a way that all readings are too large. The magnitude of the error has been found to vary in a linear fashion from 1 V for a dial reading of 5 V to 4 V for a dial reading of 80 V.
- (a) What would be the value of the voltage for each of the following dial readings: 80, 100, 50, 1, 35, 10?
- (b) Calculate the percentage error for each of the foregoing readings.

4.2 Round off each of the following numbers to three significant figures:

(a) 5.237 (b) 0.82549 (c) 81.356
(d) π (e) 6.2305 (f) 0.0428
(g) 10.45 (h) 4.035

4.3 Multiply the following measurements and express the results to the correct number of significant figures:
- (a) $6.28(10^3) \times 2.712$
- (b) 43.32×0.3
- (c) 928×4.23

4.4
- (a) Is it possible to have a precise set of measurements that are not accurate? Explain.
- (b) Is it possible to have an accurate set of measurements that are not precise? Explain.

4.5
- (a) Write the Taylor series expansion for e^x about $x = 0$.
- (b) Truncate the series found in part (a) after the fourth term and approximate e^x when $x = 0.5$. Keep five significant figures.

4.6
- (a) Evaluate the percentage error incurred in the value found for e^x in Problem 4.5(b) if the "true" value for $e^{0.5}$ is 1.6487.
- (b) Use Eq. (4.3) to find the error in $e^{0.5}$ when only four terms of

the Taylor series are taken. Is the error calculated in Problem 4.6(a) acceptable? Explain.

4.7 Evaluate

$$\left(1 + \frac{2}{x}\right)^{1/4}$$

for $x = 4$. (No calculators, please!)

4.8 Develop an expression for the coefficient of volume expansion β in terms of the change in temperature T, the initial volume V_0, and the final volume V after heating. Justify any approximations used.

4.9 How does an observer know when to stop taking measurements? In other words, how much is enough?

4.10 Make a list of five examples of systematic errors and five examples of random errors. (Do not include those mentioned in the text!)

4.11 Find the mean, median, mode, and standard deviation for the data given in the accompanying table. These data comprise 100 readings of noise levels taken about 6 miles from an airport. The readings were recorded at 15-sec intervals in the late evening.

Observed decibel values, $n = 100$									
50	50	53	48	45	51	*57	*75	*85	*82
*75	*71	*65	*61	*60	*60	*55	*55	51	50
49	49	48	51	49	54	48	49	47	49
49	49	49	49	48	47	50	49	48	49
47	48	48	50	50	54	48	47	47	48
48	49	48	47	50	49	48	48	48	48
48	48	52	50	53	49	49	48	49	47
49	55	51	50	49	48	49	45	48	50
50	51	49	50	47	47	47	47	47	47
48	50	49	49	49	49	49	49	56	49

4.12 The starred numbers in the data given in Problem 4.11 represent readings recorded while an aircraft was flying overhead. Delete the 12 starred values and find the mean, median, mode, and standard deviation of the remaining 88 measurements.

4.13 Draw a histogram of the data in Problem 4.11, and then draw a continuous curve of the number of readings as a function of noise level.

4.14 We want to find the resistance R of a resistor, so we perform the following experiment. Several different currents I are passed through the resistor, and the corresponding voltage drop V is measured very precisely with a potentiometer. The currents, how-

ever, are measured with an ordinary ammeter. Thus, random errors occur in the data.

(a) Find the resistance R by plotting the data given in the accompanying table and remembering that $V = IR$. Draw the "best-fit" line by "eyeball."

(b) Check your "eyeball" value by using the method of least squares to find R. Plot your results.

I (A) (y_i)	V (V) (x_i)
0.8	10
1.1	20
2.5	30
4.2	40
4.3	50
4.7	60
5.8	70
6.4	80

PART β

APPLICATIONS

5

FREE OSCILLATIONS OF A PENDULUM

In this chapter we shall consider in detail the first of the "physical" models that are the subject of the balance of this book. This model describes the behavior of the *pendulum*. Pendulums are found in many situations, ranging from the balance weights in grandfather clocks to playground swings to the Foucault pendulums that are commonly found in museums of science. The principal result of our model analysis will be the concept of *periodic motion*, that is, movement that repeats itself at fixed intervals of time. In this sense, of course, such natural phenomena as different as the tides and the seasons are also pendulums. As we shall see in the next chapter, the phenomenon of periodicity pervades the physical and natural sciences, but it is the pendulum that is considered to be the archetypical example of periodic motion. We shall see other "linear pendulums" in the next chapter.

In addition, we will introduce into our discussion some other valuable concepts, such as conservation (preserving) and dissipation (wasting) of energy, and in this context we shall find out how to use mathematical ideas to express "physical principles." But more of this later; first we need to start with a verbal statement of "the problem," which we can then translate into the language of mathematics.

RESULTS OF AN EXPERIMENT

Our first step in trying to build a model that would describe pendulum behavior was to conduct some simple experiments to see how a pendulum behaves. We used wooden balls with lead centers as pendulum masses, the balls being hung at the ends of ordinary pieces of string.

The geometry is as shown in Figure 5.1. The balls were held at rest initially, at some angle θ_0, and then were allowed to swing freely in planar motion. We measured the *periods of vibration* of the pendulums, that is, the cycle time T_0 for a pendulum to swing through two complete arcs (from $\theta = \theta_0$ to $\theta = -\theta_0$ and back again). The periods of vibration were measured by using photoelectric cells that were in turn connected to digital counters operating with a gated pulse as follows. The counters were triggered by one passing of the swinging mass through the photoelectric device located at the lowest part of the pendulum trajectory, and then shut off by a second passing, thus providing a direct measurement of one-half of the pendulum period.

The results of the experiments are shown in Tables 5.1 and 5.2. The experiments were conducted with two different masses—237 gm and 390 gm, respectively—attached in turn to two strings of lengths equal to 276 cm and 226 cm. The results in Table 5.1 represent a matrix of results for the small (linear) oscillations of the four pendulums obtained by permuting the two masses with the two strings. (We note that each value given as a measurement result in both Table 5.1 and Table 5.2 represents an average of five measured values.) The results in Table 5.1

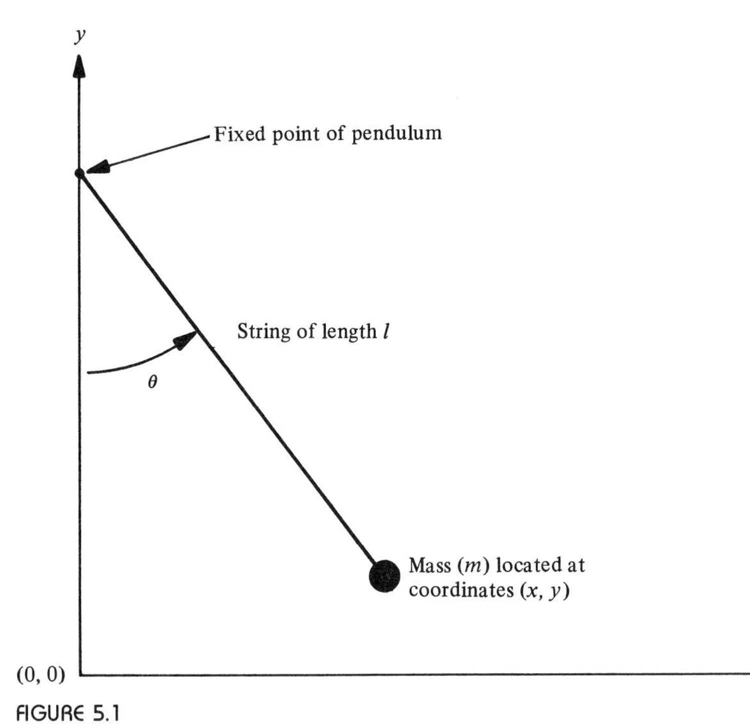

FIGURE 5.1
Geometry of a planar pendulum

Table 5.1
Periods Obtained Experimentally for Four Different Pendulums

	m = 390 gm	m = 237 gm
l = 276 cm	3.372 sec	3.350 sec
l = 226 cm	3.058 sec	3.044 sec

Table 5.2
Period Dependence on Initial Amplitude for a Pendulum[a]

θ_0 (deg)	θ_0 (rad)	T_0 measured (sec)	Measured ratio of period ($T_0/3.372$)	Calculated ratio [Eq. (5.52)]
8.34	0.1456	3.368	1.00	1.00
13.18	0.2300	3.368	1.00	1.00
18.17	0.3171	3.372	1.00	1.01
23.31	0.4068	3.372	1.00	1.01
28.71	0.5011	3.390	1.01	1.02
33.92	0.5920	3.400	1.01	1.02
39.99	0.6980	3.434	1.02	1.03
46.62	0.8137	3.462	1.03	1.04

[a]The length of the pendulum is 276 cm; T_0 for the linear pendulum case is 3.372 sec.

show that there is no appreciable variation of the period with mass, but there is a clear dependence of the period on the length of the pendulum. The results in Table 5.2 indicate that there is some slight dependence of the period of vibration on the angle from which the pendulum is allowed to swing into motion, that is, on the *initial amplitude* of the pendulum oscillation, but only at larger values of the initial amplitude (that is, $\theta_0 \geq 30°$).

Thus we conclude from the simple experiments just described that the motion of a pendulum is *periodic* (or repetitive at fixed time intervals), that the period of motion depends strongly on the length of the pendulum, and that there is a slight dependence of the period on the initial amplitude of the pendulum if the initial amplitude is fairly large.

SOME DIMENSIONAL ANALYSIS

To begin to try to make sense of the foregoing data in Table 5.1, let us do a little dimensional analysis to see if we can make something of the dependence of the period on the length of the pendulum. (We shall do some more dimensional analysis later on to extend these results.) For a start we shall see if we can use dimensional analysis to figure out precisely how the period depends on the pendulum length. Thus, we need to include as derived variables the period T_0, the string length l, and the gravitational constant g, since it is obviously gravity that makes the pendulum swing. In these three derived variables there are only two fundamental dimensions, time and length, so that either the Buckingham Pi theorem or the basic method may be readily applied. If we use the latter, we start with

$$T_0 = T_0(l, g)$$

Since the length dimension appears in a linear fashion in both l and g, it follows that

$$T_0 = T_0\left(\frac{l}{g}\right)$$

Since $[T_0]$ = time, and $[l/g] = (\text{time})^2$, it follows that

$$T_0\sqrt{\frac{g}{l}} = \text{constant} = \Pi_p \tag{5.1}$$

To determine the constant Π_p we make use of some of the data in Table 5.1. For $l = 276$ cm, one measured value of the period is 3.372 sec. With $g = 980$ cm/sec² we can use this data in Eq. (5.1) to find that

$$\Pi_p = (3.372)\sqrt{\frac{980}{276}} = 6.35 \cong 2\pi$$

In the last part of this result we have taken note of the fact that the number calculated from the data (6.35) is awfully close to a number that often pops up in analytical models ($2\pi = 6.28$). If we assume from this similarity that the period of a pendulum is in fact given by

$$T_0 = 2\pi\sqrt{\frac{l}{g}} \tag{5.2}$$

we can then calculate periods for strings of lengths used in the experiment. These results are displayed in Table 5.3. We see that the agreement with the measured values given in Table 5.1 is quite good. In fact, the differences between measured and calculated values are less than 1.5%! We can thus feel reasonably confident that, between the experi-

Table 5.3
Calculated Values of Pendulum Periods

l	226 cm	276 cm
T_0	3.02 sec	3.33 sec

mental data and the rudimentary dimensional analysis, we have a handle on some of the pendulum behavior. However, there is much more to know about the pendulum, so we need to develop some more detailed analytical models that move beyond Eq. (5.2).

EQUATIONS OF MOTION

We now formulate the problem of the pendulum as a problem in mathematical physics. Here we combine some mathematical and physical abstractions and assumptions in order to write equations that describe the free motion (without any applied forces acting) of a simple planar pendulum. This pendulum is simply a mass attached to one end of a string whose other end is fixed at a point (Figure 5.1). The mass is constrained to move in the plane of the paper, and we have chosen the coordinates (x, y) so that their origin coincides with the lowest point of the pendulum swing. The symbol m is used to denote the mass of the pendulum and l symbolizes the length of the pendulum.

The coordinates (x, y) can be expressed in terms of the string length l and the angle θ that the string makes with the ordinate. The proper expressions are

$$\begin{aligned} x &= l \sin \theta \\ y &= l(1 - \cos \theta) \end{aligned} \tag{5.3}$$

Since we intend to describe the motion of the pendulum, we shall have to account for the forces acting on the mass. That is, since the motion of the mass is predicted by *Newton's law of motion*, and since this "law" relates the coordinates (x, y) and their time rates of change to the forces on the mass, we shall need to know the forces before we can model the movement of the mass. To examine the forces in detail, consider the *free-body diagram* displayed in Figure 5.2. We shall assume that the string has so little mass of its own that it can be neglected in our model, and we shall also assume (for now) that as the mass moves through the air it is unimpeded by air resistance. Then the only forces acting on the mass are the tension T in the string and the gravitational force mg. Here, of course, g is the gravitational acceleration.

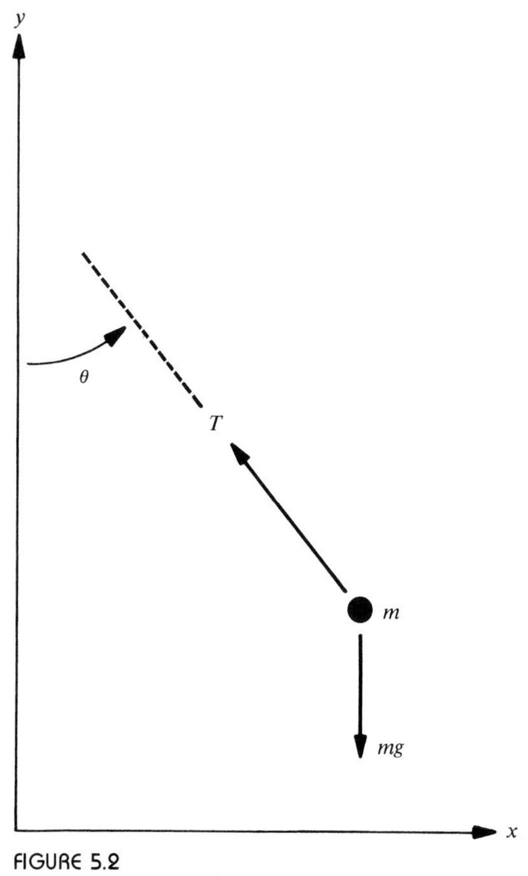

FIGURE 5.2
Forces acting on pendulum mass

Note that the tension in the string must act along the line of the string, while the gravitational force acts vertically downward, corresponding to an assumption that the y axis is roughly perpendicular to the earth's surface. Since the tension and gravitational forces are vectors and thus are assigned both a magnitude and a direction, they can be resolved into components parallel to our chosen set of axes.

Newton's law tells us that the net force on a particle causes that particle to be accelerated in direct proportion to its mass. In the symbolic language of mathematics we would write

$$\Sigma F_x = m\,\frac{d^2x}{dt^2} \qquad \Sigma F_y = m\,\frac{d^2y}{dt^2} \tag{5.4}$$

where ΣF_x and ΣF_y are the net (or resultant) forces acting on the mass in directions parallel to the x and y axes, and the terms d^2x/dt^2 and d^2y/dt^2 are the components of the acceleration of the mass parallel to the axes. The net force components can be identified from the free-body diagram (Figure 5.2) as

$$\begin{aligned} \Sigma F_x &= -T \sin \theta \\ \Sigma F_y &= T \cos \theta - mg \end{aligned} \tag{5.5}$$

As a problem in mathematics, when we wish to find the equations that describe the motion of the mass, we are saying that we want to know how the coordinates (x, y) vary with time (t). In symbolic form we can write that we are seeking the functions $x(t)$ and $y(t)$. By combining Eqs. (5.4) and (5.5) we can obtain the following pair of *differential equations of motion:*

$$\begin{aligned} m \frac{d^2x}{dt^2} &= -T \sin \theta \\ m \frac{d^2y}{dt^2} &= T \cos \theta - mg \end{aligned} \tag{5.6}$$

We are not going to solve these equations, although we shall later present solutions to a transformed set of equations that you can verify. However, it is worth noting a couple of interesting aspects of this system of equations. One point is that we do not know the string tension T that appears on the right-hand sides of Eqs. (5.6). The tension may also vary with time, but in a way that is not yet known. However, it may also be true that since we have two equations with the unknown tension, we can eliminate the unknown in a manner analogous to the elimination of unknowns in algebraic equations.

The other point of interest is that the angle θ between the string and the ordinate is also a function of time. In fact, if we knew how θ varied with time, that is, if we knew $\theta(t)$, then we would also know the coordinates of the mass because of Eqs. (5.3). If we used the relations between the coordinates, we could put our basic equations into a form where $\theta(t)$ is the only geometrical unknown. The details of this transformation of the equations are shown in the Appendix to this chapter. The transformed equations themselves are two, the first of which actually represents an equation of motion along the direction tangent to the arc of the pendulum at any instant of time:

$$ml \frac{d^2\theta}{dt^2} + mg \sin \theta = 0 \tag{5.7}$$

This is the classical differential equation that is given for the motion of a simple planar pendulum. Note that the unknown tension has been eliminated and the only unknown remaining is the angle $\theta(t)$.

A formula for the tension is the second equation derived by the transformation given in the Appendix to this chapter; it represents an equation of motion along the string when the string is at any angle θ. Here the acceleration term is the centripetal acceleration, $m(l\, d\theta/dt)^2/l$. That second equation of motion is

$$T = ml\left(\frac{d\theta}{dt}\right)^2 + mg \cos\theta \tag{5.8}$$

If we had solved Eq. (5.7) to determine $\theta(t)$, we could then determine the string tension T by a straightforward substitution into Eq. (5.8).

Before turning to the problem of analyzing the two differential equations represented by Eqs. (5.7) and (5.8), we point out again that they are equivalent equations of motion for the pendulum mass. That is, they are also a representation of Newton's law of motion, as are Eqs. (5.6). There is a difference between these two sets of equations of motion. The first set was written in reference coordinates parallel to the (x, y) axes. The second set corresponds to reference coordinates where one lies along the string and the other lies along the tangent to the path of motion of the mass. (This is the arc length θl.) We could have started out by writing Newton's law in the second set of reference coordinates and we would have gotten the same governing equations.

We also note in Eqs. (5.7) and (5.8) the presence of terms that are *nonlinear*, that is, terms where the variable $\theta(t)$ or its derivatives appear with an exponent. The most obvious of these is probably the term $(d\theta/dt)^2$ in Eq. (5.8). However, because the trigonometric functions are the series

$$\sin\theta = \theta - \theta^3/3! + \theta^5/5! - \cdots$$
$$\cos\theta = 1 - \theta^2/2! + \theta^4/4! - \cdots$$

we can readily see that these terms are nonlinear too. The nonlinear terms introduce significant complications in the solution process, so we shall defer discussion of these terms until later.

MORE DIMENSIONAL ANALYSIS

Let us now examine Eqs. (5.7) and (5.8) to see whether the dimensions are consistent. We will also look for possible simplifications that might be appropriate. In terms of the fundamental dimensions of mass, length, and time, a complete listing of the variables in the pendulum problem includes the dimensions given in Table 5.4. In view of this table it is easy to see that *all* the terms in Eqs. (5.7) and (5.8) have dimensions of force, of mass · length/(time)^2. This provides confirmation that we have done this much of the calculation correctly, since every term in an equation must have the same dimensions. Also, the dimensions of each term correspond to dimensions of force—which is obviously appropriate for an equation of motion!

We shall now introduce a *scaling factor* ω_0, which we require to have dimensions of (time)$^{-1}$. This factor in turn allows us to introduce a dimensionless time τ such that

$$\tau \equiv \omega_0 t \tag{5.9}$$

Then the differential equation (5.7) becomes (since $d/dt = \omega_0 d/d\tau$)

$$l\omega_0^2 \frac{d^2\theta(\tau)}{d\tau^2} + g \sin\theta(\tau) = 0 \tag{5.10}$$

Therefore, if we choose the scaling factor ω_0 such that

$$\omega_0 = \sqrt{g/l} \tag{5.11}$$

Eq. (5.10) immediately assumes the simple form

$$\frac{d^2\theta}{d\tau^2} + \sin\theta = 0 \tag{5.12}$$

In this (''canonical'') elegant result, all the terms are dimensionless. In view of the dimensions of l and g, we readily see from Eq. (5.11) that ω_0 does indeed have the proper dimensions. Further, reference to Eq. (5.2) indicates that the scaling factor ω_0 is related to the period T_0:

$$T_0 = 2\pi\sqrt{\frac{l}{g}} = \frac{2\pi}{\omega_0} \tag{5.13}$$

Table 5.4
Dimensions of Pendulum Variables

θ	Radians (dimensionless)
l	Length
m	Mass
t	Time
g	Length/(time)^2 = acceleration
T	Mass · length/(time)^2 = force

This relation suggests that we ought to recognize our scaling factor ω_0 as the *circular frequency* of the pendulum, that is, as the measure in radians per unit time of the periodicity of the pendulum. We shall verify this in Eq. (5.24) later in the chapter.

It is also interesting to observe in Eq. (5.12) that since $|\sin\theta| \le 1$, the acceleration term $d^2\theta/d\tau^2$ must always have a numerical value of about unity—for all possible pendulum motion, for all time! We have a time scale for the problem, too, in the following sense. If we restrict ourselves to values of dimensionless time such that $\tau \le 1$, then it will always be true that θ, $d\theta/d\tau$, and $d^2\theta/d\tau^2$ are all of roughly the same order of magnitude. For example, if

$$\theta = \theta_0 \cos\tau$$

then

$$\frac{d\theta}{d\tau} = -\theta_0 \sin\tau \qquad \frac{d^2\theta}{d\tau^2} = -\theta_0 \cos\tau$$

so that θ and its derivatives all have a maximum amplitude of θ_0. Note, however, that this does not mean that in terms of the actual physical time t that θ, $d\theta/dt$, and $d^2\theta/dt^2$ are of the same size.

Do we need to restrict τ to be about unity in size? No, we don't. From Eqs. (5.9) and (5.11) we see that

$$\tau = t\sqrt{g/l} = \frac{t}{\sqrt{l/g}} \tag{5.14}$$

In slightly different terms, Eq. (5.14) can be written as a "verbal equation,"

$$\tau = \frac{\text{actual (physical) time}}{\text{some constant with units of time}}$$

If we expected the process to take years, we would pick the constant in the denominator to be in terms of years. In that case, if τ were very small, we would be looking at an actual time period in hours or days. If τ became very large, say much greater than unity, it might mean that we ought to choose the constant as a number expressed in decades or even centuries.

Conversely, as seems to be the case with the pendulum, the physics of the problem—or, a particular problem—may dictate the time scale. For example, for a pendulum that is 2 ft long,

$$\omega_0 = \sqrt{g/l} \cong 4 \text{ sec}^{-1}$$

so that we could say that a *characteristic time* for the system is approximately $\frac{1}{4}$ sec; that is, we expect the repetitive behavior of the system to occur in a short period of time. For a very long pendulum, say $l = 200$ ft, we would have $\omega_0 \cong 0.40 \text{ sec}^{-1}$, or a characteristic time of about $2\frac{1}{2}$ sec.

LINEAR MODEL OF THE PENDULUM

We have already observed that the pendulum equation [Eq. (5.7)] is nonlinear in nature. Nevertheless, it is a problem whose solution can be obtained analytically. Here, however, we are interested in developing the linear approximation to the pendulum, the classical *simple harmonic oscillator*. Let us assume that the angle of the pendulum can be written in the form

$$\theta(\tau) = \theta_0 f(\tau) \tag{5.15}$$

where, *by assumption*,

$$\theta_0 = \max|\theta(\tau)| \tag{5.16}$$

so that $|f(\tau)| \leq 1$. Then we can regard θ_0 as the *amplitude* of motion, as the correct indicator of the magnitude of the pendulum swings. If θ_0 is very small, the pendulum arcs over a small angle. Conversely, if θ_0 is quite large, we expect big swings. Note that the terms "small" and "large" are very vague and imprecise, since we have not indicated small or large with respect to anything in particular.

We showed in Chapter 4 that even for fairly large angles (that is, $-30° \leq \theta \leq 30°$) we can take

$$\begin{aligned} \sin\theta &\cong \theta \qquad \text{percentage error } 4.51\% \\ \cos\theta &\cong 1 \qquad \text{percentage error } 15.5\% \end{aligned} \tag{5.17}$$

If we assume, in Eq. (5.15), that $|\theta_0| \leq 30°$, we can simplify Eq. (5.12) to the result

$$\frac{d^2\theta}{d\tau^2} + \theta = 0 \tag{5.18}$$

Further, the tension in the pendulum can then be found from Eq. (5.8) as

$$T = mg\left[1 + \left(\frac{d\theta}{d\tau}\right)^2\right] \tag{5.19}$$

wherein we have used the second approximation of Eqs. (5.17). Is this result, Eq. (5.19), consistent with the approximations made in Eq. (5.17)?

In fact, Eq. (5.19) is not an approximation that is consistent with Eqs. (5.17) and (5.18). The question of consistency comes to mind because of the presence in the bracketed expression of Eq. (5.19) of the nonlinear term $(d\theta/d\tau)^2$ added to unity. Since $d\theta/d\tau$ must be of the same order of magnitude as θ [viz., Eq. (5.15) and the discussion on pp. 94–95], and since we assume $\theta^2 << 1$ in writing $\cos\theta = 1$, it would obviously be inconsistent to keep the nonlinear term in Eq. (5.19). Therefore, for the linear model, we have the interesting result that

$$T \cong mg \tag{5.20}$$

Thus, even for swings of the pendulum out to $\pm 30°$, the tension in the string is nearly constant!

Now we will turn our attention to solving Eq. (5.18), by which we mean that we are looking for a function $\theta(\tau)$ that will identically satisfy Eq. (5.18). As a trial solution* we consider the function

$$\theta(\tau) = \theta_0 \cos\tau \tag{5.21}$$

which is also of the form of Eq. (5.15). Since we can show that

$$\frac{d^2\theta}{d\tau^2} = -\theta_0 \cos\tau = -\theta$$

it is obvious that this solution does satisfy the differential equation (5.18). The solution does have at least one flaw, however, since we don't know the magnitude of θ_0, the amplitude.

We can notice, though, that at time $t = \tau = 0$ we have from Eq. (5.21) the result that $\theta(0) = \theta_0$. Thus θ_0 represents the *initial* amplitude of motion of the pendulum. We may also notice that the solution (5.21) is such that at $t = \tau = 0$, $d\theta(0)/d\tau = 0$; that is, the initial speed of the pendulum is zero. Thus this corresponds to initially holding the pendulum at rest at an arbitrary angle θ_0 and then letting it go.

Another solution that satisfies the linearized equation of motion is

$$\theta(\tau) = \dot{\theta}_0 \sin\tau \tag{5.22}$$

You can readily verify that this corresponds to giving the mass an

*Wherever a solution for a differential equation is required, it will be presented and you will be asked only to verify it.

initial angular velocity $\dot{\theta}_0$ at the origin [as $\theta(0) = 0$]. Then if each of Eqs. (5.21) and (5.22) satisfies the differential equation, what about their sum? In fact the sum does satisfy the equation and, since θ_0 and $\dot{\theta}_0$ can be chosen independently, we can use this result to describe the motion of a pendulum that is given an arbitrary initial angular velocity $d\theta/d\tau$ at some arbitrary angle. Thus a general solution* to the pendulum equation is

$$\theta(\tau) = \theta_0 \cos \tau + \dot{\theta}_0 \sin \tau \tag{5.23}$$

You will recall from trigonometry that the functions $\cos \tau$, $\sin \tau$ are periodic; that is, if their arguments are increased by 2π, the same value of the function is repeated. Therefore, when $\tau = 2\pi$, $\theta(2\pi) = \theta(0)$. Thus, in physical time, the value of $\theta(t)$ repeats at time intervals such that

$$t = nT_0 = \frac{2\pi n}{\omega_0}, \qquad n = 1, 2, 3, \ldots \tag{5.24}$$

Hence the identification of T_0 as the period of motion and ω_0 as the angular frequency—which has the units of radians per unit of time. We can also define a frequency $f_0 = 1/T_0 = \omega_0/2\pi$ with units of $(\text{time})^{-1}$. We have also, incidentally, verified Eqs. (5.2) and (5.13).

A final observation is in order before turning to other considerations. We note from the discussion surrounding Eqs. (5.23) and (5.24) and the definition of ω_0 [Eq. (5.11)] that the period of vibration depends only on physical characteristics of the pendulum and *not* on the amplitude of the motion. This is an important characteristic of linear vibration and dynamics problems, as is the superposition of solutions as given in Eq. (5.23). That is, two hallmarks of linear problems are the ability to add solutions and the independence (or uncoupling) of the amplitude of the motion from the period.

ENERGY CONSIDERATIONS—CONSERVATION

Now we shall see what we can learn from examining the energy of the ideal system where there is no dissipation or waste of energy. The total energy is the sum of the kinetic and potential energies. The kinetic

*The theory of linear differential equations indicates that there will be two independent solutions for a second-order equation, which is an equation whose highest-order derivative is the second.

energy can be calculated as

$$
\begin{aligned}
\mathrm{KE} &= \frac{1}{2}m(\text{speed})^2 \\
&= \frac{1}{2}m\left[\left(\frac{dx}{dt}\right)^2 + \left(\frac{dy}{dt}\right)^2\right] \\
&= \frac{1}{2}m\left[\left(l\cos\theta\,\frac{d\theta}{dt}\right)^2 + \left(l\sin\theta\,\frac{d\theta}{dt}\right)^2\right]
\end{aligned}
$$

or

$$
\mathrm{KE} = \frac{1}{2}ml^2\left(\frac{d\theta}{dt}\right)^2 = \frac{1}{2}mgl\left(\frac{d\theta}{d\tau}\right)^2 \tag{5.25}
$$

The potential energy of the mass, with the origin of the coordinate system ($x = 0$, $y = 0$) taken as the datum, is

$$
\begin{aligned}
\mathrm{PE} &= mg\cdot(\text{distance above the datum}) \\
&= mgy = mgl(1 - \cos\theta)
\end{aligned} \tag{5.26}
$$

For the linear pendulum model, we should use the trigonometric expansion for the term $\cos\theta$ in Eq. (5.26) and, since we are looking for the difference between unity and $\cos\theta$, we have to note that

$$
\begin{aligned}
1 - \cos\theta &= 1 - [1 - \theta^2/2! + \theta^4/4! - \cdots] \\
&= \theta^2/2 + \cdots
\end{aligned} \tag{5.27}
$$

Again, as we have discussed in Chapter 4, we cannot simply take $\cos\theta \cong 1$ in this context, since that leaves us with a trivial and silly—and incorrect—result. Thus, for the linear case

$$
\mathrm{PE} = \frac{1}{2}mgl\,\theta^2 \tag{5.28}
$$

The total energy is then

$$
\begin{aligned}
E &= \mathrm{KE} + \mathrm{PE} \\
&= \frac{1}{2}mgl\left[\left(\frac{d\theta}{d\tau}\right)^2 + \theta^2\right] = E(\tau)
\end{aligned} \tag{5.29}
$$

Note that the dimensions of energy are force · distance, which are the dimensions of work.

How does the energy vary with time? Indeed, does the energy change with time? To see, let us simply calculate the time rate of change from Eq. (5.29):

$$\frac{dE(\tau)}{d\tau} = \frac{1}{2}mgl\left(2\,\frac{d\theta}{d\tau}\,\frac{d^2\theta}{d\tau^2} + 2\theta\,\frac{d\theta}{d\tau}\right)$$

$$= mgl\left(\frac{d^2\theta}{d\tau^2} + \theta\right)\frac{d\theta}{d\tau} \tag{5.30}$$

This is a remarkable result! When we look at the term in parentheses and compare it to Eq. (5.18), the equation of motion, we see that the equation of motion is reproduced in the calculation of the time rate of change of energy. Further, if $\theta(\tau)$ is such that the equation of motion is identically satisfied, then

$$\frac{dE(\tau)}{d\tau} = 0 \quad \text{and} \quad E(\tau) = E_0 = \text{constant} \tag{5.31}$$

Thus, energy is conserved!

If the energy is constant, we may go on to ask: What is this constant value? There are two ways to answer this question, and we shall present the answers in terms of the solution (5.21) for the pendulum released from rest at some arbitrary initial angle. The first answer derives from simply substituting that solution into the equation for the total energy, Eq. (5.29):

$$E_0 = \frac{1}{2}mgl[(-\theta_0 \sin \tau)^2 + (\theta_0 \cos \tau)^2]$$

$$= \frac{1}{2}mgl\theta_0^2 \tag{5.31a}$$

A second way to answer the question is to say that since the energy is constant, that constant value must be the energy of the mass when the motion starts. Initially, for the solution (5.21), the mass has only potential energy, since it is released from rest at some arbitrary angle. The initial potential energy is, from Eq. (5.28),

$$\text{PE}(\tau = 0) = \frac{1}{2}mgl\theta_0^2 \equiv E_0 \tag{5.31b}$$

The energy approach developed in the preceding discussion can also be applied to the nonlinear pendulum model and, as we shall see, it can be used to provide a hint as to how to find $\theta(\tau)$ to solve the nonlinear problem. From Eqs. (5.25) and (5.26), using $\tilde{E}(\tau)$ to denote the energy in the nonlinear model, we have

$$\tilde{E}(\tau) = mgl \left[\frac{1}{2}\left(\frac{d\theta}{d\tau}\right)^2 + 1 - \cos\theta \right] \tag{5.32}$$

whose time derivative is

$$\frac{d\tilde{E}(\tau)}{d\tau} = mgl \left(\frac{d^2\theta}{d\tau^2} + \sin\theta \right) \frac{d\theta}{d\tau} \tag{5.33}$$

Clearly, if $\theta(\tau)$ satisfies the nonlinear differential equation for the pendulum, then the time rate of change of energy in the pendulum is zero, and energy is conserved.

For both the linear and nonlinear models of the pendulum, we have seen that by differentiating the energy we have produced equations of the form

$$\frac{d(\text{total energy})}{d\tau} = (\text{exact diff. eq.}) \frac{d\theta}{d\tau}$$

which has generally turned out to be zero. This suggests—as we shall see in detail in a later section—that there may be a way to start integrating some of these complex differential equations by *reversing* the process. That is, we would try to construct integrals of the form

$$[\text{diff. eq.}]\left(\frac{d\theta}{d\tau}\right) \cdot d\tau = ?$$

However, more of this in the section after next, where we shall construct an "energy integral" for the nonlinear pendulum model.

ENERGY CONSIDERATIONS—DISSIPATION

In the preceding section we considered ideal pendulums wherein no energy was wasted. Here we will examine the consequences of including in the differential equation a *dissipative force*. Such forces arise because of resistance to motion of a body through air, for example, or because of frictional resistance to rotation in a less-than-perfect joint. The dissipative force, often called a *friction force* or a *damping force*, is generally assumed to be proportional to the speed of the object being analyzed, the constant of proportionality being called a *coefficient of friction*. For a *viscous friction force*, we can write that

$$\text{friction force} = -R(\text{velocity}) \tag{5.34}$$

where R is a positive constant with dimensions of force per unit speed. The minus sign is introduced here to acknowledge that frictional forces act to slow down the motion by opposing it. In the nonlinear pendulum equation, then, the inclusion of a damping force changes Eq. (5.7) into

$$ml \frac{d^2\theta}{dt^2} + Rl \frac{d\theta}{dt} + mg \sin \theta = 0 \tag{5.35}$$

What are the effects of this damping force on the energy in the system? The kinetic and potential energies are unchanged (in form), so that we can use Eqs. (5.25) and (5.26) to write the total energy—using now the actual physical time t—as

$$E(t) = \frac{1}{2} ml^2 \left(\frac{d\theta}{dt}\right)^2 + mgl(1 - \cos \theta) \tag{5.36}$$

If we calculate the time rate of change of $E(t)$, we have

$$\frac{dE(t)}{dt} = \left[ml^2 \frac{d^2\theta}{dt^2} + mgl \sin \theta\right] \frac{d\theta}{dt}$$

or, in view of Eq. (5.35),

$$\frac{dE(t)}{dt} = -Rl^2 \left(\frac{d\theta}{dt}\right)^2 \tag{5.37}$$

Therefore, we see that the energy in the system is continually decreasing with time, since $(d\theta/dt)^2 \geq 0$.

We can take this result a step further by arguing that, in some sort of average sense, the values of kinetic and potential energy are equal to each other. This was obviously the case when we looked at the motion from rest of a linear pendulum. It also looks like a reasonable argument from Eq. (5.29), since the time scaling there is done so that $\theta(\tau)$, $d\theta/d\tau$, etc., are all of the same magnitude. If we assume then, as a rough but good approximation, that average kinetic energy = average potential energy, it follows that on average

$$E(t) \cong 2 \cdot (\text{kinetic energy})$$

or

$$E(t) \cong ml^2 \left(\frac{d\theta}{dt}\right)^2 \tag{5.38}$$

And, by combining Eqs. (5.37) and (5.38) to eliminate the term $(d\theta/dt)^2$, we wind up with a single differential equation for the energy as a function of time:

$$\frac{dE(t)}{dt} = -(R/m)E(t) \tag{5.39}$$

Note that the dimensions of R/m are force/(velocity · mass) or, in terms of the basic dimensions, $(\text{time})^{-1}$. Thus both sides of Eq. (5.39) correctly have the same dimensions.

From the theory of differential equations we can easily solve the differential equation (5.39). The solution is, as you can easily verify,

$$E(t) = E_0 e^{-Rt/m} \tag{5.40}$$

where $E_0 = E(t = 0)$ is the starting value of the energy. Thus the energy decays exponentially from its starting value, which is also its maximum value. The rate of energy decay depends on the coefficient of friction R, and a *characteristic time* of energy decay could be expressed in terms of the ratio m/R. This ratio has the dimension of time, and its precise value in terms of seconds, days, or centuries would depend on the particular pendulum being analyzed. However, it is easy to construct Table 5.5, which shows the energy decay. Note that the energy is

Table 5.5
Decay of Oscillator Energy with Time

Time	Energy
$t = 0$	$E(t) = E_0$
$= 0.10(m/R)$	$= 0.905E_0$
$= 0.69(m/R)$	$= 0.500E_0$
$= 1.00(m/R)$	$= 0.368E_0$
$= 5.00(m/R)$	$= 0.007E_0$

halved in a time equal to $0.69(m/R)$. This could serve as a useful indicator of energy decay time.

A final note: This entire set of manipulations on energy conservation and dissipation was conducted without knowing the form of $\theta(t)$. That is, we have obtained much significant information by looking at—but not solving—the equation of motion of the pendulum and by examining such physically based notions as energy, without knowing $\theta(t)$! Thus we can obtain information about a system without necessarily solving a differential equation, and simultaneously we obtain general results that are valid for a class of problems.

NONLINEAR MODEL OF THE PENDULUM*

We indicated earlier that the process of examining the time rate of change of energy, in which the equation of motion is produced, could perhaps be inverted to produce "integrals" or "solutions" to the equation of motion. That is, perhaps by starting with the differential equation and the knowledge of the other process, we could find functions $\theta(t)$ to satisfy the nonlinear pendulum equation, Eq. (5.12). Some of the mathematics will be complex and considerably beyond what has been required, but the results will be of interest even if some of the details are skipped. We begin with Eq. (5.12) written in the form

$$\left(\frac{d^2\theta}{d\tau^2} + \sin\theta\right)\frac{d\theta}{d\tau} = 0$$

which we can also recognize as

$$\frac{d}{d\tau}\left[\frac{1}{2}\left(\frac{d\theta}{d\tau}\right)^2 - \cos\theta\right] = 0 \tag{5.41}$$

Note the resemblance of the bracketed term of Eq. (5.41) to the energy expressions we have used before. Again, inverting the differentiation of the energy, it is clear from Eq. (5.41) that

$$\left[\frac{1}{2}\left(\frac{d\theta}{d\tau}\right)^2 - \cos\theta\right] \equiv \text{constant} \tag{5.42}$$

If we were to consider the case of the pendulum started at rest from an

*This section, save for the result in Eq. (5.52), can be skipped since it is not central to the discussion.

arbitrary angle θ_0, then at $\tau = 0$

$$\theta(\tau) = \theta_0, \qquad \frac{d\theta(\tau)}{d\tau} = 0 \tag{5.43}$$

Clearly, since the left-hand side of Eq. (5.42) is a constant, and since the conditions of Eqs. (5.43) must apply, that constant must be $-\cos\theta_0$; that is,

$$\frac{1}{2}\left(\frac{d\theta}{d\tau}\right)^2 - \cos\theta = -\cos\theta_0$$

or

$$\left(\frac{d\theta}{d\tau}\right)^2 = 2(\cos\theta - \cos\theta_0) \tag{5.44}$$

Now it is useful to write the trigonometric identity

$$\cos\theta = 1 - 2\sin^2\frac{\theta}{2}$$

so that we can rewrite Eq. (5.44) in the form

$$\left(\frac{d\theta}{d\tau}\right)^2 = 4\left(\sin^2\frac{\theta_0}{2} - \sin^2\frac{\theta}{2}\right)$$

or, taking a square root on both sides and rearranging terms,

$$2\,d\tau = \frac{d\theta}{\sqrt{\sin^2\dfrac{\theta_0}{2} - \sin^2\dfrac{\theta}{2}}} \tag{5.45}$$

We shall now integrate Eq. (5.45) over one quarter of a full period of pendulum motion. Over a complete period, $0 \le t \le \tilde{T}_0$, where $\tilde{T}_0$ is the period for the nonlinear model, the pendulum swings from $\theta = -\theta_0$ to $\theta = +\theta_0$ and back again, both times passing through the "origin," the coordinate $\theta = 0$. Note that for the nonlinear problem, while we can define $\tilde{T}_0$ as a period of motion, and can use dimensionless time as defined by $\tau = \omega_0 t = t\sqrt{g/l}$, we *cannot* make the identification $\tilde{T}_0 = 2\pi/\omega_0$. A quarter-period would be the time interval $0 \le \tau \le \pi\tilde{T}_0/2T_0$ or $0 \le t \le \tilde{T}_0/4$, during which the pendulum swings from $\theta = -\theta_0$ to $\theta = 0$. Expressing this integration symbolically means writing the integral of Eq. (5.45):

$$2\int_0^{\pi T_0/2\tilde{T}_0} d\tau = \int_{-\theta_0}^{0} \frac{d\theta}{\sqrt{\sin^2\dfrac{\theta_0}{2} - \sin^2\dfrac{\theta}{2}}} \tag{5.46}$$

Or, in a slightly different form with the easy integration on the left carried out,

$$\frac{\pi \tilde{T}_0}{T_0} = \int_0^{\theta_0} \frac{d\theta}{\sqrt{\sin^2 \frac{\theta_0}{2} - \sin^2 \frac{\theta}{2}}} \tag{5.47}$$

Equation (5.47) contains the *implicit solution* to the pendulum problem evaluated for fixed limits of t and θ. For variable limits, the corresponding integral is called an implicit solution because instead of finding $\theta(t)$ explicitly, we have a solution that will produce values of time for given values of the angle θ. It also turns out that although the integral on the right-hand side of Eq. (5.47) is messy looking, it is a standard integral, called an *elliptic integral*, for which there are published tables of numerical values for given values of the constant θ_0.

However, we can perform a few more manipulations to get another nice result. Introduce a new angle ϕ defined by the relation

$$\sin \frac{\theta}{2} = \sin \frac{\theta_0}{2} \sin \phi \tag{5.48}$$

The range of values of ϕ is such that $0 \leq \phi \leq \pi/2$ for $0 \leq \theta \leq \theta_0$. Next, define a new constant p such that

$$p^2 = \sin^2 \frac{\theta_0}{2} \tag{5.49}$$

From Eq. (5.48) it follows by differentiation that

$$\frac{1}{2} \cos \frac{\theta}{2} \, d\theta = \sin \frac{\theta_0}{2} \cos \phi \, d\phi$$

or

$$\begin{aligned} d\theta &= \frac{2p \cos \phi}{\cos \frac{\theta}{2}} \, d\phi \\ &= 2p \frac{\sqrt{1 - \sin^2 \phi}}{\sqrt{1 - \sin^2 \frac{\theta}{2}}} \, d\phi \\ &= 2p \frac{\sqrt{1 - \sin^2 \phi}}{\sqrt{1 - p^2 \sin^2 \phi}} \, d\phi \end{aligned} \tag{5.50}$$

Then the integral in Eq. (5.47) can be written as

$$\pi \frac{\tilde{T}_0}{T_0} = \int_0^{\pi/2} \frac{2p\sqrt{1 - \sin^2 \phi}\, d\phi}{\sqrt{1 - p^2 \sin^2 \phi}\, \sqrt{p^2 - p^2 \sin^2 \phi}}$$

or

$$\frac{\tilde{T}_0}{T_0} = \frac{2}{\pi} \int_0^{\pi/2} \frac{d\phi}{\sqrt{1 - p^2 \sin^2 \phi}} \tag{5.51}$$

Equation (5.51) is a neater and cleaner version of the elliptic integral we mentioned earlier. The tables of numerical values of the integral—as a function of $p = \sin(\theta_0/2)$—can be used to construct another table or a curve of the relationship between $\tilde{T}_0$ and p. In other words, since p is dependent on the magnitude of the initial swing angle θ_0, the solution to the full nonlinear problem demonstrates a dependence of the period $\tilde{T}_0$ on the angle θ_0. If we neglected $p^2 \sin^2 \phi$ compared to unity in the denominator of the integral—an approximation of the same type as that used in developing the linear pendulum model [Eqs. (5.13), (5.17), and (5.18)]—we would recoup the period of the linear pendulum, $T_0 = 2\pi/\omega_0$, which is independent of the amplitude θ_0.

An intermediate calculation can be made in order to make this last set of ideas more explicit. It we used the binomial expansion developed in Chapter 4, we could write

$$(1 - p^2 \sin^2 \phi)^{-1/2} \cong 1 + \frac{1}{2} p^2 \sin^2 \phi$$

Then the expression for the period, Eq. (5.51), becomes

$$\begin{aligned} \frac{\tilde{T}_0}{T_0} &\cong \frac{2}{\pi} \int_0^{\pi/2} \left(1 + \frac{1}{2} p^2 \sin^2 \phi\right) d\phi \\ &\cong \frac{2}{\pi} \left[\frac{\pi}{2} + \frac{1}{2} p^2 \int_0^{\pi/2} \sin^2 \phi\, d\phi\right] \end{aligned}$$

The integral is now of a standard form that is independent of the parameter p. Evaluating the integral, we find

$$\begin{aligned} \frac{\tilde{T}_0}{T_0} &\cong \left(1 + \frac{1}{4} p^2\right) \\ &\cong \left(1 + \frac{1}{4} \sin^2 \frac{\theta_0}{2}\right) \\ &\cong \left(1 + \frac{1}{16} \theta_0^2\right) \end{aligned} \tag{5.52}$$

where in writing Eq. (5.52) we have again used the small-angle approximation for the sinusoid. Clearly and explicitly, the period $\tilde{T}_0$ depends on the amplitude θ_0 of the pendulum motion. We can also see that Eq. (5.52) is a ratio of a result from the nonlinear model to the corresponding result from the linear model. We showed measured and calculated values of this ratio in Table 5.2. Again, as we have with various other aspects of this problem, we see very good agreement between the empirical data and the predictions from the model.

SUMMARY

In this chapter we have presented an extended discussion of the motion of a pendulum in order to show: (1) how a mathematical model of a physical problem is derived; (2) how it can be checked for consistency; and (3) what kind of information can be derived from the model. We have used the concepts of scaling, we have checked for dimensional consistency, and we have developed several approximations. We have also demonstrated how well our model of the pendulum predicts (or confirms) actual physical (laboratory) behavior.

In terms of the pendulum itself, we have shown how the period in a linear model of the pendulum depends only on the physical properties of the pendulum and not on the motion. This is a hallmark of linear problems. We have also developed an explicit dependence of the period on amplitude for the nonlinear model of the pendulum, and have related this result back to the linear model.

Problems

5.1 (a) Use Eq. (5.52) to determine the maximum value of θ_0 if the ratio T_0(nonlinear)/T_0(linear) may not exceed 1.005.

(b) Estimate the value of $\theta_{0_{max}}$ that could be employed with satisfactory accuracy before a correction factor should be used.

5.2 (a) From the data given in the accompanying table, determine what

		Period of Revolution (sec)					
θ	m	$l_1 = 1$ m			$l_2 = 3$ m		
θ_1	m_1	2.09	2.09	2.10	3.45	3.40	3.48
	m_2	2.07	2.08	2.08	3.46	3.44	3.44
θ_2	m_1	1.95	1.98	1.94	3.37	3.40	3.38
	m_2	1.96	1.93	1.95	3.36	3.38	3.35
θ_3	m_1	1.87	1.87	1.88	3.24	3.29	3.27
	m_2	1.86	1.85	1.87	3.22	3.25	3.21

variables affect the period of the conical pendulum shown in the accompanying figure ($\theta_1 < \theta_2 < \theta_3$ and $m_1 < m_2$).

(b) Use dimensional analysis to show how the period depends on these variables.

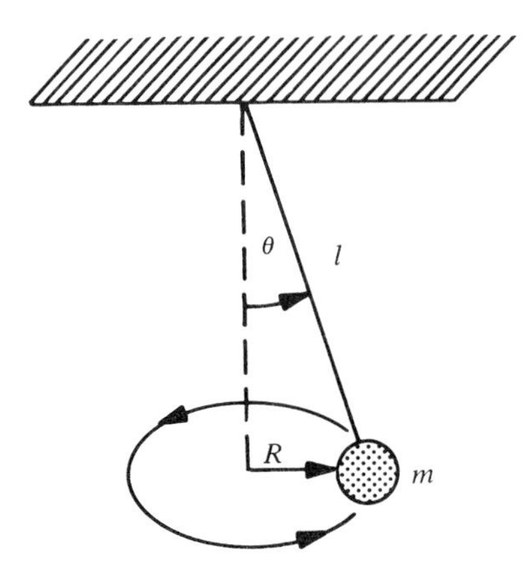

5.3 Check the answer to Problem 5.2(b) by deriving the period of the conical pendulum from the equations of motion.

5.4 A uniform meter stick is supported at one end and is allowed to swing about a pivot through that end. The period of oscillation of this stick differs from that of a simple pendulum since the mass is distributed along the length l of the "pendulum" rather than concentrated at a single point. Use dimensional analysis to show how the period of this pendulum depends on the pendulum mass per unit length m, the length l, and the gravitational constant g.

5.5 For the meter stick in Problem 5.4, find the complete equation for the period of oscillation for small oscillations; that is, find the value of the unknown constant(s). Use Newton's laws of rotational motion to write an equation from which an analogy to the simple pendulum case may be used. The rotational inertia (moment of inertia) I about the pivot point is $\frac{1}{3}mL^2$.

5.6 If we take the kinetic energy of the pendulum in Problem 5.5 as $\frac{1}{2}I(d\theta/dt)^2$, and the potential energy as the weight times the height of the center of mass above its lowest position, show that the total energy of the system is conserved.

5.7 (a) Find the rate at which energy is being dissipated for a simple planar pendulum when the damping force is proportional to the (velocity)^2.

(b) Show that the answer in Problem 5.7(a) is dimensionally correct.

5.8 (a) Assuming that there is no dissipation of energy in a vibrating spring-mass sytem (Figure 2.3), write an equation for the total energy of the system as a function of the spring constant k and the maximum displacement $A(x_{max})$ from the unstretched spring position. Take the potential energy stored in the spring as PE $= \frac{1}{2}kx^2$ where x is the displacement from the unstretched position. The spring and mass m are vibrating horizontally.

(b) Show that the answer in Problem 5.8(a) is dimensionally correct.

5.9 Kepler's third law of planetary motion can be written as an equation for the square of the periods of revolution about the sun:

$$T^2 = \frac{4\pi^2 a^3}{GM_s}$$

Here a is the semimajor axis of the elliptical planetary orbit, M_s is the mass of the sun, and G is the universal gravitational constant.

(a) Starting with this equation, find an equation for the frequency in the form $\omega = f(G, M_s, a)$.

(b) The force of attraction between a planet and the sun is

$$F = \frac{GM_s(\text{mass of planet})}{(\text{distance between planet and sun})^2}$$

What approximation is needed to derive the equation of motion of Problem 5.9(a)? Explain the answer in mathematical terms.

5.10 Is energy conserved in planetary motion? Explain. [The gravitational potential energy is $GM_s m_{planet}$/(distance between sun and planet).]

5.11 Show from Eq. (5.7) that the mass of a simple pendulum achieves its maximum velocity at $\theta = 0°$. Is this physically reasonable?

5.12 Show that the result of Problem 5.11 obtains for both the linear and nonlinear models of the pendulum.

5.13 In the actual experiments with pendulums, would you expect the energy to remain conserved? If so, why? If not, how would you expect to see effects of energy dissipation?

APPENDIX: TRANSFORMATION OF EQUATIONS OF MOTION*

We display here the steps in the transformation of Eqs. (5.6) into Eqs. (5.7) and (5.8). What we wish to do first is eliminate the terms $m\, d^2x/dt^2$ and $m\, d^2y/dt^2$ in the equations

$$m \frac{d^2x}{dt^2} = -T \sin \theta$$
$$m \frac{d^2y}{dt^2} = T \cos \theta - mg \tag{5.6}$$

To eliminate these terms we first apply the chain rule of the calculus:

$$\frac{dx}{dt} = \frac{dx}{d\theta}\frac{d\theta}{dt} \tag{A.1}$$

You may then verify by repeated application of Eq. (A.1) to Eqs. (5.3) that

$$\frac{d^2x}{dt^2} = l \cos \theta \frac{d^2\theta}{dt^2} - l \sin \theta \left(\frac{d\theta}{dt}\right)^2$$
$$\frac{d^2y}{dt^2} = l \sin \theta \frac{d^2\theta}{dt^2} + l \cos \theta \left(\frac{d\theta}{dt}\right)^2 \tag{A.2}$$

Thus, by this bit of mathematical manipulation we have come to a point where the coordinates (x, y) can be eliminated from Eqs. (5.6) with Eqs. (A.2):

$$ml \cos \theta \frac{d^2\theta}{dt^2} + \left[T - ml \left(\frac{d\theta}{dt}\right)^2\right] \sin \theta = 0$$
$$ml \sin \theta \frac{d^2\theta}{dt^2} - \left[T - ml \left(\frac{d\theta}{dt}\right)^2\right] \cos \theta = -mg \tag{A.3}$$

It is not clear, looking at this formidable set of equations, that we have improved the situation! However, observe that if we multiply the first equation by $\cos \theta$, the second by $\sin \theta$, and then add the two equations—exactly as we would with algebraic equations—then the

*This section can be skipped; it is not central to the discussion.

bracketed term will fall out:

$$ml(\cos^2 \theta + \sin^2 \theta) \frac{d^2\theta}{dt^2} = -mg \sin \theta$$

or

$$ml \frac{d^2\theta}{dt^2} + mg \sin \theta = 0 \tag{5.7}$$

Again we have thus found the classical equation, in $\theta(t)$ alone, for the pendulum, the equation of motion along the tangent to the pendulum arc. A formula for the tension can be determined in a manner similar to the calculation just performed. If we multiply the first of Eqs. (A.3) by $\sin \theta$, the second by $\cos \theta$, and then take the difference, we find that

$$\left[T - ml \left(\frac{d\theta}{dt}\right)^2\right] (\sin^2 \theta + \cos^2 \theta) = mg \cos \theta$$

or

$$T = ml \left(\frac{d\theta}{dt}\right)^2 + mg \cos \theta \tag{5.8}$$

This is the second equation of motion, in a direction along the string, from which the tension can be calculated once $\theta(t)$ has been determined from Eq. (5.7).

6

FORCED MOTION OF LINEAR OSCILLATORS

The purpose of this chapter is to expand the discussion of the linear pendulum model of Chapter 5 in order to develop more physical models whose mathematical descriptions are identical. We shall now look beyond periodic oscillation where no external force is present and examine the "forced" motion of simple periodic systems. We shall show that the simple harmonic oscillator has wide applicability in the physical sciences; consequently, it is worth a fair amount of attention.

THE SPRING-MASS OSCILLATOR

We begin with the system shown in Figure 6.1, that system consisting of a mass being pulled on one side by a force $F(t)$ that is dependent only on time, and being pulled on the other side by an *elastic spring*. The elastic spring is an ideal, massless spring that always exerts a restoring force on the mass to return it to its unstretched condition. The restoring force is directly proportional to the distance the mass has been moved, the constant of proportionality being the *spring stiffness*, which has units of force per unit displacement. The motion of the mass will be denoted by the coordinate $x(t)$, measured again from the unstretched ("equilibrium") position of the spring.

We can now apply Newton's law to the mass; this law states that the net force on the mass causes it to accelerate according to the relation

$$\text{net force} = m\,\frac{d^2x}{dt^2} \qquad (6.1)$$

The net force on the mass is the resultant of the applied force $F(t)$

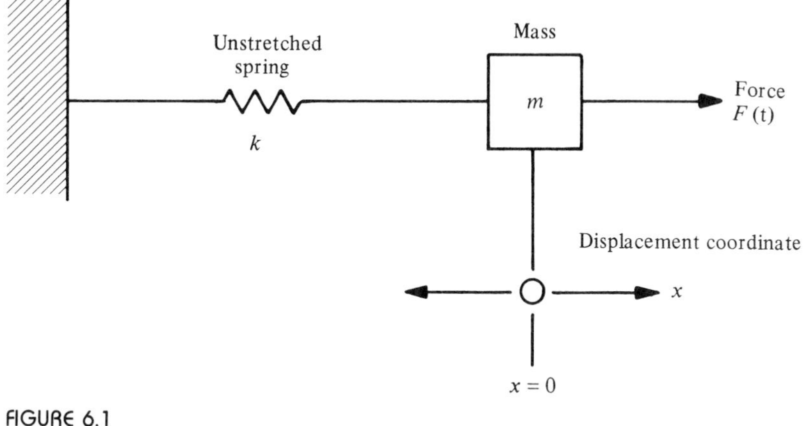

FIGURE 6.1
Elementary spring-mass system

acting to the right in the drawing and the spring restoring force $-kx$, which would act to the left if the mass were displaced by a positive distance $x(t)$ to the right. Hence,

$$\text{net force} = F(t) - kx \tag{6.2}$$

so that by combining Eqs. (6.1) and (6.2) we have the equation of motion for the mass in the mass-spring system:

$$m \frac{d^2x}{dt^2} + kx = F(t) \tag{6.3}$$

This is the classical equation of the *simple harmonic oscillator*, often simply referred to as the *linear oscillator*.

Equation (6.3) has exactly the same form as the linear pendulum model [see Eq. (5.18)], which implies that the restoring force for the pendulum is the tangential component of the gravitational pull on the pendulum mass. This makes physical sense, for if we push on a pendulum mass in the θ direction, it can only be gravity that pulls the mass back down to the lowest (equilibrium) position. Of course, in our previous work on the pendulum we considered only free oscillation, so that there was no corresponding forcing term; that is, there was no $F(t)$.

We have also considered the damped pendulum, that is, a pendulum whose motion was retarded by air resistance or some other dissipative mechanism. There is an analogous spring-mass-damper system, shown in Figure 6.2. Here the damper exerts a resistive force on the

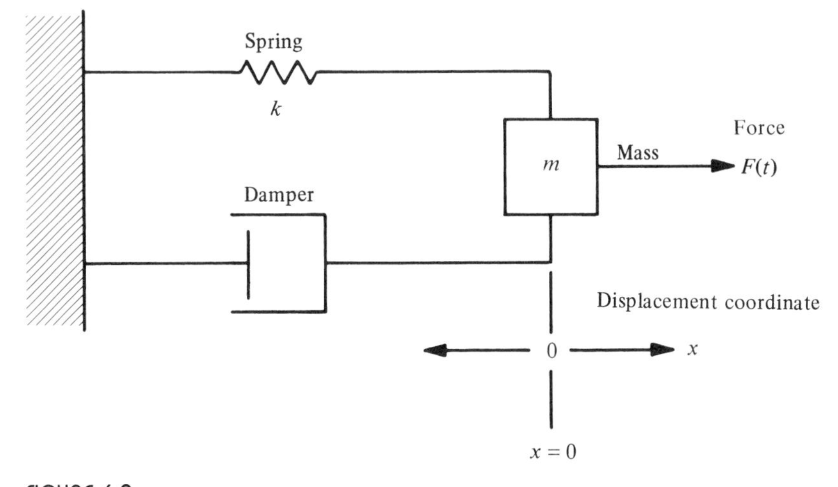

FIGURE 6.2
Spring–mass–damper system

mass proportional to the speed of the mass $dx(t)/dt$. The constant of proportionality is again denoted by R and has dimensions of force per unit velocity. The net force on the mass is then

$$\text{net force} = F(t) - kx - R\frac{dx}{dt} \tag{6.4}$$

so that the appropriate equation of motion is

$$m\frac{d^2x}{dt^2} + R\frac{dx}{dt} + kx = F(t) \tag{6.5}$$

Again this formula can be compared to the corresponding pendulum result as given in Eq. (5.35)—without the small-angle approximation for the pendulum sinusoid.

We note here a point (to which we shall return later) about the spring, mass, and damping elements: We observe that the damper was inserted as a resistive or dissipative device. We have seen already [Eq. (5.40)] that the presence of a damper reduces the total energy in the system. Hence, the damper is referred to as an *energy-dissipation element*. In a similar way, the spring and the mass are referred to as *energy-storing elements* because they store energy as potential energy and kinetic energy, respectively. They may only store the energy put into them for a very short time, particularly for high-frequency vibration, but whatever is stored can be recovered; none of it is lost!

Later in this chapter we shall present a more complete picture of the solutions to the linear oscillator equation, after we have presented several physical problems that can be described analytically by the oscillator equation. Before proceeding to the examples, however, we want to repeat from the preceding chapter some of the scaling ideas relevant to the oscillator.

In particular, for free oscillations, we again introduce a dimensionless time τ, as in Eq. (5.9). If this is done in Eq. (6.3) with $F(t) = 0$, we find that the oscillator equation becomes

$$\frac{d^2x}{d\tau^2} + x = 0 \tag{6.6}$$

if the constant ω_0 is such that

$$\omega_0 = \sqrt{\frac{k}{m}} \tag{6.7}$$

Since this result is very different—at least in appearance—from the corresponding pendulum result of Eq. (5.11), we should immediately check the dimensions. The dimensions of m are clear enough, while for the stiffness k we have

$$[k] = \frac{\text{force}}{\text{length}} = \frac{\text{mass} \cdot \text{length}}{\text{length} \cdot (\text{time})^2} = \frac{\text{mass}}{(\text{time})^2}$$

so that the dimensions of ω_0 are correct; that is

$$[\omega_0] = \sqrt{\frac{\text{mass}}{\text{mass}(\text{time})^2}} = (\text{time})^{-1}$$

As before, we can identify ω_0 as the *circular frequency* of the system, and we can also relate it to the period and the frequency

$$f_0 = \frac{1}{T_0} = \frac{\omega_0}{2\pi} = \frac{1}{2\pi}\sqrt{\frac{k}{m}} \tag{6.8}$$

Note that both ω_0 and f_0 have physical dimensions of $(\text{time})^{-1}$, but the former has to be expressed in radians per unit time, the latter in cycles (radians divided by 2π) per unit time.

We note in Eqs. (6.7) and (6.8) that the frequency is proportional to the ratio of the stiffness to the mass, or more specifically to the square root of that ratio. In the pendulum problem we have

$$\omega_0 = \sqrt{\frac{g}{l}} = \sqrt{\frac{mg}{ml}} = \sqrt{\frac{mg/l}{m}} \tag{6.9}$$

Recall that we spoke of the restoring force of the pendulum as being the gravitational force mg. In Eq. (6.9) we see the gravitational force being expressed as a stiffness: the gravitational force mg divided by the length l of the pendulum.

This is a pattern that we shall see many times: The frequency of an oscillator is proportional to the square root of the stiffness-to-mass ratio. Hence, the stiffer the system, the shorter its period. Also, if we reduce the mass of a system, we reduce the period or, equivalently, we increase the frequency. We shall have many occasions to note this relation because changing the frequency (or period) of a system is often a useful design objective.

BUILDING VIBRATION

It is not surprising that buildings, especially tall and slender skyscrapers, vibrate in the presence of several kinds of forces. The buildings respond to aerodynamic forces due to wind as well as to groundborne vibration which can be caused by traffic, explosions, or even earthquakes. Further, these external sources of vibration, which may cause the whole building to oscillate as does a cattail in the wind, may also cause internal components (such as floors, windows, or walls) to vibrate, and this internal vibration can create problems of annoyance or dysfunction, if not outright danger, for the building occupants. The assessment of the vibration problems of a new or an existing building requires some understanding of the dynamic characteristics of the building, some understanding of the building's natural period. A "first-order" simple estimate can be obtained by modeling the building as a simple mass–spring system.

Consider the building shown in Figure 6.3. We have drawn a simple tall building with a uniformly distributed wind pressure acting on one building face along the entire height of the skyscraper. The wind pressure tends to push the building, with the result that the building bends to yield the largest movement at the top (the free end). Since a building is made up of elastic structural members, we expect it to resist the bending action of the wind and to restore itself to a straight vertical position when no wind is acting. In this sense we can visualize the building acting like the reed or beam of the idealized model in Figure 6.4. But even this model can be simplified—if an appropriate stiffness and mass can be identified—into the spring–mass system of Figure 6.5. The question is: What are appropriate values of k and m to characterize the building properly?

At least a couple of elementary approaches come to mind. One is to

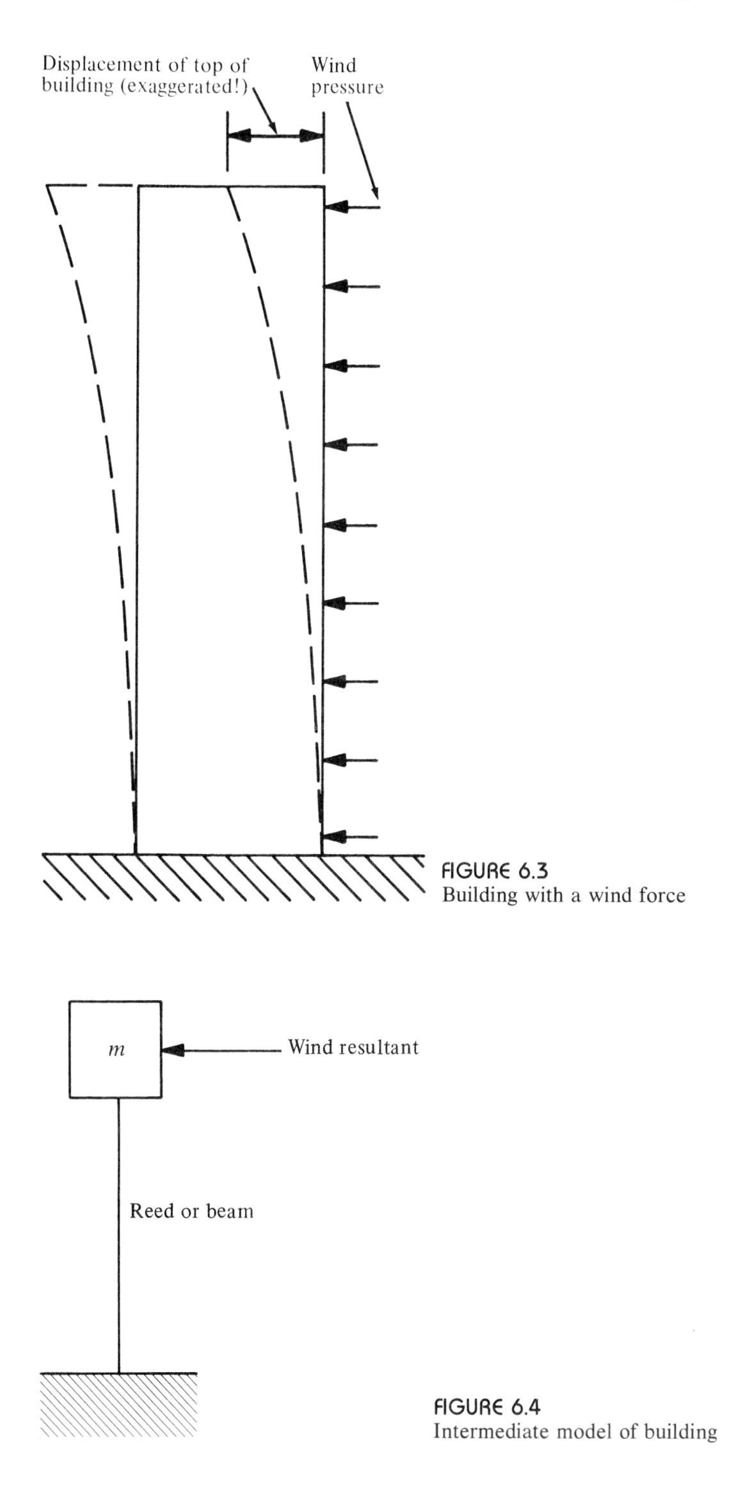

FIGURE 6.3
Building with a wind force

FIGURE 6.4
Intermediate model of building

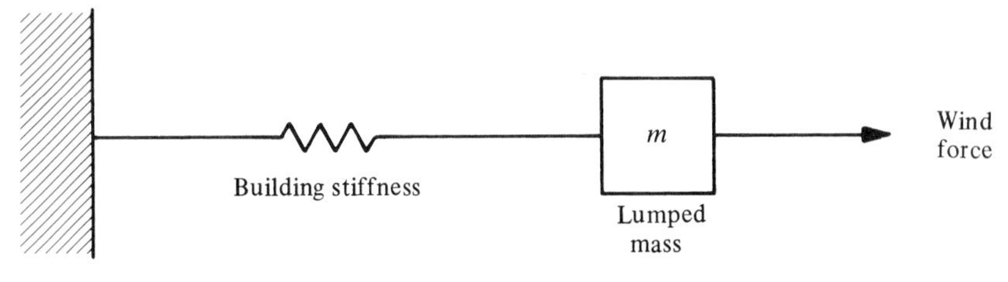

FIGURE 6.5
Representation of building as a spring–mass system

look at how much the top of the building moves under a given wind loading and then use this information to back-calculate the stiffness (the mass is presumed known in the form of the weight of the structure). For example, a building some 1000 ft (300 m) high with a square cross section of 100 ft on a side (30 m by 30 m) can be expected to deflect as much as 1.5 in. (0.038 m) in a gale-force wind. Such a wind could exert a pressure of 14 lb/ft² (670 N/m²) on a side of the building. The total wind force is

$$\text{wind force} = \begin{cases} 14 \text{ lb/ft}^2 \times 100 \text{ ft} \times 1000 \text{ ft} \\ 670 \text{ N/m}^2 \times 30 \text{ m} \times 300 \text{ m} \end{cases} = \begin{cases} 1.4 \times 10^6 \text{ lb} \\ 6.0 \times 10^6 \text{ N} \end{cases}$$

If we assume this force is concentrated at the mid-height of the building, and that this point moves one half of the top movement, or 0.75 in. (0.019 m), then we can calculate a stiffness for the building as

$$k_{\text{building}} = \begin{cases} 1{,}400{,}000 \text{ lb/0.75 in.} \\ 6{,}000{,}000 \text{ N/0.019 m} \end{cases} = \begin{cases} 1.87 \times 10^6 \text{ lb/in.} \\ 3.16 \times 10^8 \text{ N/m} \end{cases}$$

The building weight is obtained from the structural engineer, who assigns it the value of 2.20×10^{11} lb (about 9.8×10^{11} N). This corresponds to a mass of 6.85×10^9 slugs (10^{11} kg). Then the natural frequency is

$$f_0 = \frac{1}{2\pi}\sqrt{\frac{3.16 \times 10^8 \text{ N/m}}{10^{11} \text{ kg}}}$$
$$= 0.0089 \text{ cycles/sec}$$

which is quite low. This corresponds to a period of $T_0 = 112.4$ sec.

Another way of estimating the stiffness of the building is to treat it as a cantilever beam held in an upright position, as in Figure 6.4. In this instance we can use the theory of strength of materials to show that the stiffness of such a beam is given by

$$k_{\text{cantilever}} = \frac{3EI}{L^3} \tag{6.10}$$

where E is the modulus of elasticity of the material of which the beam is made (with dimensions of force per unit area), I is the moment of inertia of the cross-sectional area of the beam [with dimensions of $(\text{length})^4$], and L is the length of the beam (with dimensions of length). It is easily verified that $k_{\text{cantilever}}$ has dimensions of force per unit length, as it should. We can also see that the result in Eq. (6.10) is intuitively pleasing in terms of what it says about a building's stiffness. To explore this intuition, a couple of facts about E and I will be useful.

The modulus E represents the stiffness of the material itself, and the modulus of steel is greater than that of aluminum, which is greater than that of wood, and so on. So, in very loose terms, we can view a higher modulus as being more suitable for use in buildings because of the greater material stiffness implied. The moment of inertia I represents the distribution of material in a beam cross section about the centerline of that cross section. The greater the thickness of the beam (or building), the greater—in general—the value of I is. Thus a building whose plan area is 30 m × 30 m would have a greater moment of inertia than a building that measures 25 m × 25 m.

We then see from Eq. (6.10) that a short, squat building will be much stiffer than a tall, slender building. Two buildings of the same material and plan area will have their stiffnesses related to each other by the cube of the inverse ratio of their heights. Thus the taller one will have a lower stiffness, and will therefore be more flexible and have a lower natural frequency.

This explains why engineers have had to worry about the effects of wind on buildings only in the last couple of decades, since the advent of high-strength steels and new design approaches that have made possible the construction of flexible skyscrapers. The estimation of the frequencies of buildings has now become a standard part of the design process in order that wind loading, earthquakes, and other sources of dynamic loading or stress, can be properly considered.

AUTOMOBILE SUSPENSION SYSTEM

An automobile suspension system, from our everyday practical experience, is an obvious mass-spring-damper system. The mass is clearly reflected in the mass of the auto and its cargo; there are springs at all four wheels—whether coil springs or leaf springs; and the shock absorbers provide the damping mechanism. What is new for us at this point is that the auto is not being pushed or pulled by any obvious external force. Consider Figure 6.6, which shows the subject automobile as it rides along a roadway to be a single mass-spring-damper system.* Here the excitation—or the external stimulus—comes from the fact that the roadway is causing the lower end of the mass-spring-damper system to move up and down. We may visualize this system as a modification of the system originally presented in Figure 6.2. The modification depicted in Figure 6.7 shows motion of the spring-damper at the left side to reflect the imposed deformation due to the wheel acting on the road. Further, both the spring and the damper are shown as acting in *parallel* in that the relative motion between the two ends of each element is the same. Therefore the relative extension of the spring is $x(t) - a(t)$, while the relative speed to which the damper responds is

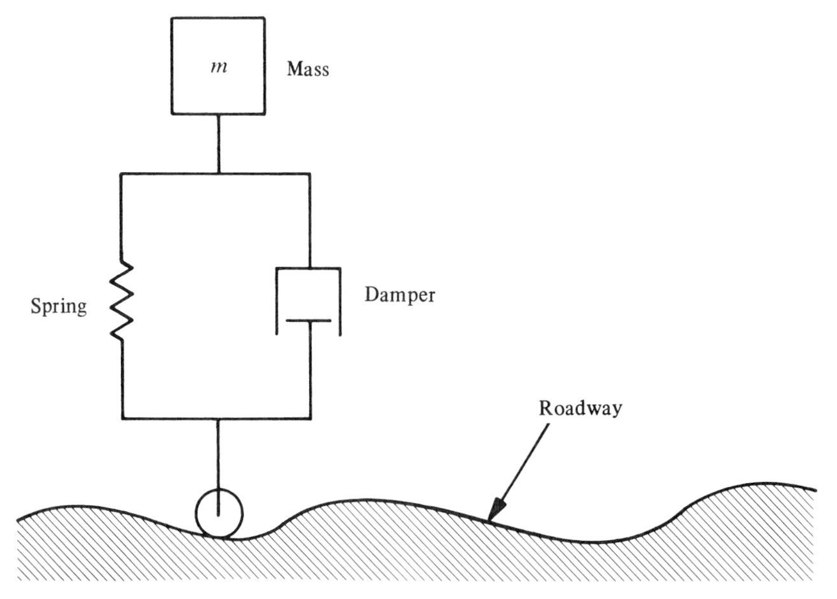

FIGURE 6.6
Schematic automobile on roadway

*We show this schematic in a way that conforms with our experience as to how shock absorbers and springs are mounted together in a car with a common free end at the junction of wheel and axle.

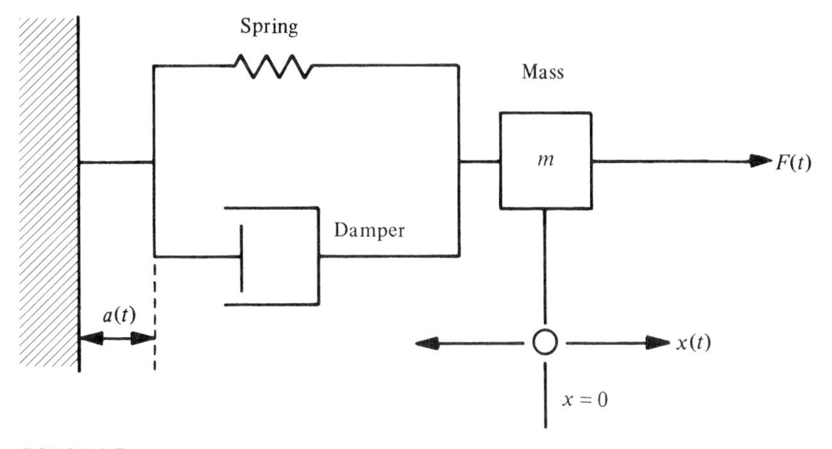

FIGURE 6.7
Mass-spring-damper system with moving support

$d[x(t) - a(t)]/dt$. The spring force is then $k(x - a)$ while the damper force is $R(dx/dt - da/dt)$, so that Newton's law for this system is

$$m\frac{d^2x}{dt^2} = F(t) - k(x - a) - R\left(\frac{dx}{dt} - \frac{da}{dt}\right)$$

or

$$m\frac{d^2x}{dt^2} + R\frac{dx}{dt} + kx = F(t) + ka(t) + R\frac{da(t)}{dt} \tag{6.11}$$

In Eq. (6.11) we have written the terms involving the movement of the foundation $a(t)$ on the right-hand side—as a known function of time.

As a simple example, let there be no external forces acting on our auto, so that $F(t) = 0$. Let the auto be traveling along the sinusoidal path shown in Figure 6.8. The sinusoidal path is taken as

$$y(z) = a_0 \sin \alpha z \tag{6.12}$$

where α is a constant with dimensions of $(\text{distance})^{-1}$. If the automobile moves down the road at constant speed v, it follows that $z = vt$. Then the foundation movement of the model is found to be

$$a(t) = y(z = vt) = a_0 \sin \alpha vt \tag{6.13}$$

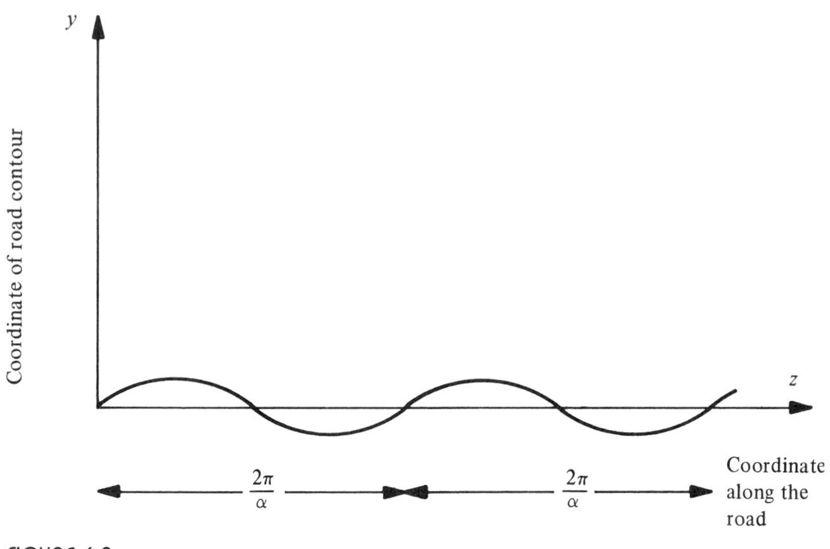

FIGURE 6.8
Sinusoidal roadway

The spring-mass-damper system response is the solution of

$$m \frac{d^2x}{dt^2} + R \frac{dx}{dt} + kx = ka_0 \sin \alpha vt + R\alpha va_0 \cos \alpha vt \tag{6.14}$$

ACOUSTIC RESONATOR

The pastime of blowing across the open top of a bottle to produce a foghorn type of sound is well known. Indeed, the pitch (or frequency) of the resulting sound is known to vary with the amount of liquid in the bottle or, more precisely, with the size of the air cavity in the bottle. A flask or bottle with an air space that is used to produce sound is called an *acoustic resonator* or a *Helmholtz resonator*, after the German physicist who investigated it. Consider the flask shown in Figure 6.9. There is an "interior" cavity of volume V_0 that contains a gas of density ρ_0 at some ambient pressure p_0. The neck of the flask has length l and a cross-sectional area A. We shall see that it is the gas in the neck that moves, with the cavity providing a springlike resistance to that motion.

If we develop such an analogy, we can take the mass of the spring-mass system as $m = \rho_0 Al$. The stiffness exerted by the air cavity will

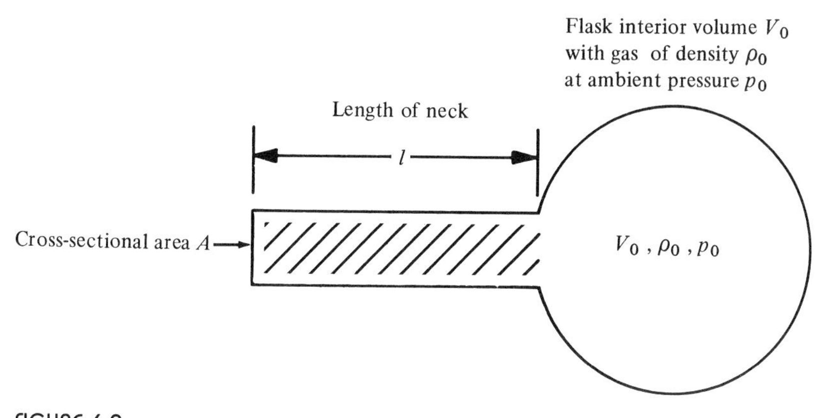

FIGURE 6.9
Acoustic (Helmholtz) resonator

have to be obtained from the pressure p_0 at the neck-cavity interface. The pressure p_0 can in turn be calculated from the adiabatic gas law, which the gas in the cavity is presumed to obey. That law is

$$pV^{\gamma} = \text{constant} \tag{6.15}$$

where γ is the ratio of the specific heats,* and p and V are pressure and volume, respectively. When the mass of gas in the neck moves to the right by a distance x, the volume of gas in the cavity must be decreased by an amount

$$\delta V = -Ax \tag{6.16}$$

From the adiabatic gas law, if we "differentiate" both sides of Eq. (6.15), we find that

$$\delta p V^{\gamma} + \gamma V^{\gamma-1} p \, \delta V = 0$$

or dividing through by pV^{γ},

$$\frac{\delta p}{p} + \gamma \frac{\delta V}{V} = 0$$

and letting the pressure and volume take on their ambient values:

$$\delta p = -\gamma \frac{p_0}{V_0} \delta V \tag{6.17}$$

*For air, $\gamma = 1.4$.

Now, by combining Eqs. (6.17) and (6.16) to eliminate the volume change, we can relate the pressure change δp to the distance x that the mass moves:

$$\delta p = \gamma \frac{p_0}{V_0} A x$$

This result represents a spring stiffness because the restoring force acting on the mass would be

$$F_{\text{restoring}} = A\,\delta p = \gamma \frac{p_0 A^2}{V_0} x \tag{6.18}$$

Therefore, if our blowing across the bottle produces a dynamic force $F(t)$, the mass $m = \rho_0 A l$ vibrates against the stiffness $k = \gamma p_0 A^2/V_0$ in accordance with the oscillator equation:

$$\rho_0 A l \frac{d^2x}{dt^2} + \gamma \frac{p_0 A^2}{V_0} x = F(t) \tag{6.19}$$

We can rewrite this result in terms of the speed of sound c_0 of the gas, which is related to the specific heat and the ambient pressure and density by the formula

$$c_0^2 = \gamma \frac{p_0}{\rho_0} \tag{6.20}$$

Thus our oscillator equation (6.19) becomes

$$\rho_0 A l \frac{d^2x}{dt^2} + \frac{\rho_0 c_0^2 A^2}{V_0} x = F(t) \tag{6.21}$$

It is worth checking the dimensions of this stiffness term, for it certainly does not look like a stiffness. However, we can note that

$$\begin{aligned}\left[\frac{\rho_0 c_0^2 A^2}{V_0}\right] &= \frac{\text{mass}}{(\text{length})^3} \; \frac{(\text{length/time})^2(\text{length})^4}{(\text{length})^3} \\ &= \frac{\text{mass}}{(\text{time})^2} \equiv \frac{\text{force}}{\text{length}} = \text{stiffness}\end{aligned}$$

Further, we note that the acoustic stiffness term is

$$k_{\text{acoustic}} = \frac{\rho_0 c_0^2 A^2}{V_0} = \frac{\gamma p_0 A^2}{V_0} \tag{6.22}$$

which shows that if the gas is contained at a higher ambient pressure, then the stiffness will increase. If the flask volume increases, the stiffness will decrease, and the flask will therefore be more compliant. Both of those effects are intuitively pleasing. Finally, as we did in Eq. (6.7), we can calculate the frequency of this system as

$$\omega_0 = c_0 \sqrt{\frac{A}{lV_0}} \tag{6.23}$$

Hence, the larger the flask volume, the lower the frequency, and the deeper the foghorn sound!

PARTICLE MOVING IN A MAGNETIC FIELD*

Whenever a charged particle such as an electron, a proton, or an ion moves through a magnetic field, that particle is subject to a magnetic force. That force is given by the formula

$$\mathbf{F}_m = q\mathbf{v} \times \mathbf{B} \tag{6.24}$$

where **B** is the vector representing the *magnetic induction* due to currents other than that produced by the charge q, **v** is the velocity vector of the charged particle, and q is the magnitude of the charge of the particle. In the mks system of units the unit of charge is the coulomb. The magnetic induction is also called the *magnetic flux density* and it has units in the mks electromagnetic system of *webers per square meter* (Wb/m^2), which can also be expressed as kilograms per coulomb-second (kg/C-sec). Further, to relate this to force units of newtons, we use the equation

$$\text{newtons} = \text{coulombs} \times \frac{\text{meters}}{\text{second}} \times \frac{\text{kilograms}}{\text{coulomb-seconds}}$$
$$= \frac{\text{kilograms} \times \text{meters}}{(\text{seconds})^2}$$

The symbol × in Eq. (6.24) denotes a vector cross product, which indicates that the resultant force $\mathbf{F}_m$ lies perpendicular to the plane of the two vectors **v** and **B**. This relationship is depicted in Figure 6.10.

*This section requires some knowledge of vectors and vector products.

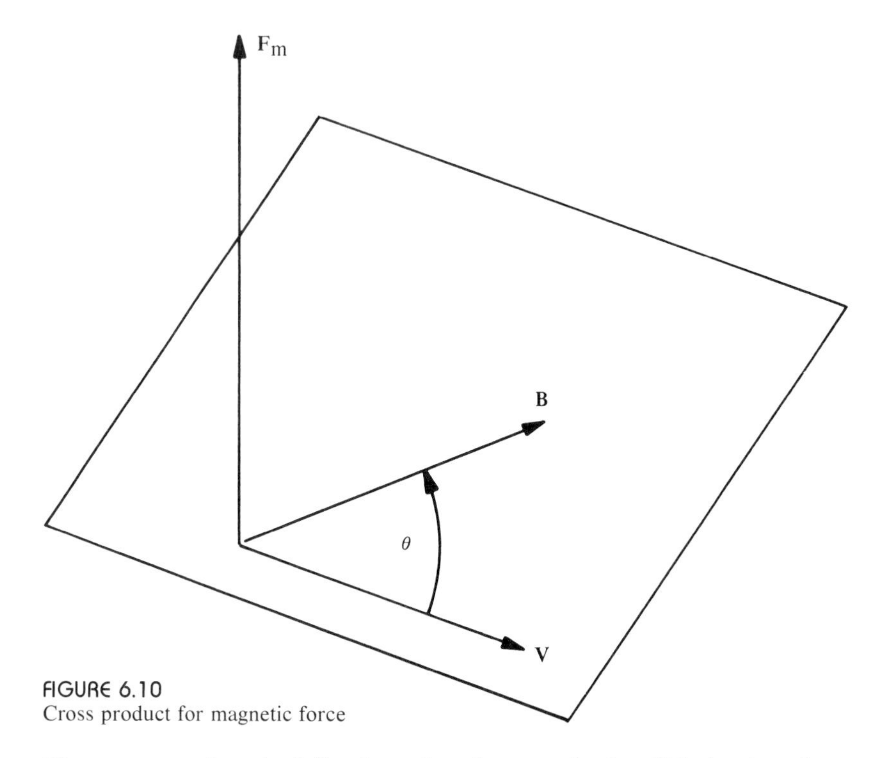

FIGURE 6.10
Cross product for magnetic force

The cross product is defined so that the magnitude of $\mathbf{F}_m$ is given by

$$|\mathbf{F}_m| = q|\mathbf{v}||\mathbf{B}|\sin\theta \tag{6.25}$$

where θ is the angle between the two vectors $\mathbf{v}$ and $\mathbf{B}$.

We note from Eq. (6.24) that, by the definition of the cross product, the force on the particle is always perpendicular to the particle velocity—assuming of course that the magnetic field does not change with time. Thus, the magnetic field imparts no energy to the particle, and since the force and acceleration are normal to the velocity, the particle must be traveling in a circle. This is already a clue that we will be looking at simple harmonic motion, for that motion in the form $x = x_0 \cos \omega_0 t$ is also a description of the x coordinate of a particle moving around a circle of radius x_0 with a radian frequency ω_0.

Now, because the force and acceleration are perpendicular to the velocity, and because the particle travels in a circle with the velocity vector being tangent to the circle, it follows that the acceleration is directed radially inward. In fact, we find from mechanics that the radial acceleration is $r\omega^2$,* where r is the radius of the circle and ω is the

*This acceleration can also be written in terms of a tangential velocity component v_T as v_T^2/r because of Eq. (6.27).

radian frequency of the motion. Thus applying Newton's law in the radial direction

$$\Sigma F = mr\omega^2 \tag{6.26}$$

together with the fact that the tangential velocity is

$$v_T = \omega r \tag{6.27}$$

and with Eq. (6.25), we have

$$mr\omega^2 = qv_T B = q\omega r B$$

Then the frequency of the oscillation of a charged particle in a constant magnetic field is simply

$$\omega = \frac{q}{m} B \tag{6.28}$$

This result, called the *cyclotron frequency*, shows that the frequency depends only on the strength of the magnetic flux density B and on the charge-to-mass ratio q/m of the particle. It is independent of the radius of the circle, and thus it is also not dependent on the tangential velocity v_T. Equation (6.28) is the fundamental relation behind the design of the cyclotrons, from whence it derives its name.

ELECTRICAL CIRCUITS AND THE ELECTRICAL-MECHANICAL ANALOGY

As a final example of the linear oscillator we will consider a simple electrical circuit. We go to this example partly to extend the range of application of the same mathematical model and partly to develop one of the most widely used analogies in science. This is the *electrical-mechanical analogy* and it is of great utility in such fields as acoustics and analysis of mechanical vibrations. We begin by considering the three electrical elements shown in Figure 6.11.

Sketched in Figure 6.11 are three *passive circuit elements* that produce voltage responses that depend in different ways on the flow of charge (and its derivatives) through them. These are idealized elements in the same way that in the spring-mass system the mass is assumed to be rigid and the spring is assumed to have no mass of its own. The first element is the *capacitor*, which upon discharge transmits a voltage that

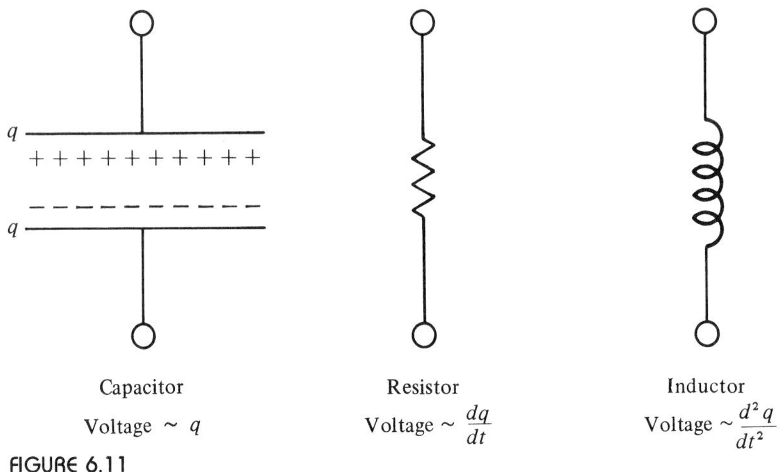

FIGURE 6.11
Passive electric circuit elements

is proportional to the charge stored on two plates that are separated by an insulator. The relationship between the voltage and the charge is given by

$$V_C = q/C \tag{6.29a}$$

where C is the *capacitance*.*

The second element is a *resistor*, which impedes the flow of electrical current I in proportion to the current, which is itself defined as the time rate of change of the charge. Hence, the voltage drop across a resistor is given by *Ohm's law*:

$$V_R = RI = R\frac{dq}{dt} \tag{6.29b}$$

Here the coefficient of proportionality is the *resistance* R, and the *current* is $I \equiv dq/dt$.

The third element is an *inductor*. This is a coil that builds up a magnetic field within itself when there is a current flowing through it. The magnetic field causes a voltage difference to develop between the ends of the coil, that voltage being proportional to the time rate of

*We shall for the sake of simplicity defer introduction of units until the electrical–mechanical analogy is developed.

change of the current. Hence the voltage drop across an inductor is

$$V_{\mathrm{L}} = L\,\frac{dI}{dt} = L\,\frac{d^2q}{dt^2} \tag{6.29c}$$

The coefficient L is the *self-inductance* of the coil.

Note from Eqs. (6.29) that we have three elements whose responses (voltages) are directly proportional to the charge, the first derivative of the charge with respect to time, and the second time derivative, respectively. Thus, if we treated the charge q like the displacement of an oscillator, the capacitor would act like a spring in that the voltage is proportional to q. The resistor then would appear as a damper, and the inductor as a mass. We shall continue this line of reasoning a bit later. For now let us turn to the electrical circuit displayed in Figure 6.12.

We show in that figure a simple electrical circuit, called an *RLC circuit*, which consists of a loop containing a capacitor, a resistor, an inductor, and a source of an applied voltage $V_a(t)$. In order to analyze this circuit we will make use of Kirchhoff's voltage law.* The voltage law is a statement of equilibrium for the circuit, analogous to a balance of forces: "The sum of all voltage drops and inputs around any closed path must be zero at every instant." In symbolic form, for the circuit in Figure 6.12, we have then

$$V_{\mathrm{a}}(t) = V_{\mathrm{C}} + V_{\mathrm{R}} + V_{\mathrm{L}} \tag{6.30}$$

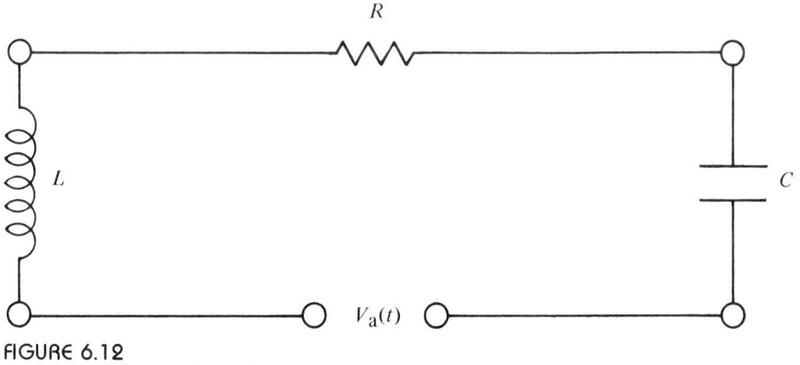

FIGURE 6.12
Elementary electrical oscillation circuit

*There is also a Kirchhoff current law, which is analogous to a statement of conservation of mass. If in a circuit there are multiple loops that connect at various *nodes*, the current law states that the sum of all currents entering (+) and leaving (−) a node must be zero.

If we substitute from Eqs. (6.29) for the voltages in the elements, we find a very familiar equation:

$$L \frac{d^2q}{dt^2} + R \frac{dq}{dt} + \frac{1}{C} q = V_a(t) \tag{6.31}$$

This is identical in form to Eq. (6.5), which is the equation governing the oscillations of the spring-mass-damper system. The construction of Table 6.1 is now rather straightforward. (We have included proper units for the mks system.)

FORCED MOTION OF A LINEAR OSCILLATOR: RESONANCE AND IMPEDANCE

We have now seen several instances of the role that the linear harmonic oscillator plays in describing important and interesting physical problems. This part of the discussion will be devoted to a discussion of the solution of the oscillator equation when it is subject to a periodic forcing function $F_0 \cos \omega t$. Thus, we shall consider the following mathematical problem. We wish to find $x(t)$ such that

$$m \frac{d^2x}{dt^2} + kx = F_0 \cos \omega t \tag{6.32}$$

Here the frequency ω may be prescribed at will, it is arbitrary and is called either the *forcing frequency* or the *driving frequency*. It is not the

Table 6.1
Electrical-Mechanical Analogy

Mechanical quantity (units)	Electrical quantity (units)
Force (newtons)	Voltage (volts)
Displacement (meters)	Charge (coulombs)
Velocity (meters/second)	Current (amperes)
Mass (kilograms)	Inductance (henrys)
Compliance or (stiffness)$^{-1}$ (meters/newton)	Capacitance (farads)
Viscous friction (force/velocity)	Resistance (ohms)
Natural frequency ($\omega_0^2 = k/m$)	Natural frequency ($\omega_0^2 = 1/LC$)
Period ($T_0 = 2\pi\sqrt{m/k}$)	Period ($T_0 = 2\pi\sqrt{LC}$)

same as the natural frequency ω_0 of the spring-mass system, although it may assume that value.

The complete solution to Eq. (6.32) contains two types of terms. The first type of solution is periodic with a frequency equal to the natural frequency ω_0 and it reflects the *initial conditions* as in Eq. (5.23). In terms of the dimensional time t,

$$x_t(t) = C_1 \cos \omega_0 t + C_2 \sin \omega_0 t \tag{6.33}$$

where C_1 and C_2 are functions of $x(0)$ and $dx(0)/dt$. We have denoted this solution as $x_t(t)$. It is usually called the *transient solution* because it depends on the initial conditions. The other type of solution is the *steady-state response*, and you may verify by differentiation that the steady-state solution is

$$x_s(t) = \frac{F_0}{m(\omega_0^2 - \omega^2)} \cos \omega t, \qquad \omega \neq \omega_0 \tag{6.34}$$

A complete solution to the linear equation (6.32) is given by the *sum** of the solutions in Eqs. (6.33) and (6.34).

Our interest is confined to the steady-state solution, for it predicts the response of the spring-mass system to a periodic input for all time. Further, it is independent of the initial conditions, which can always be incorporated by appropriate choice of the constants in Eq. (6.33). So we take as our solution

$$x(t) = \frac{F_0}{m(\omega_0^2 - \omega^2)} \cos \omega t \tag{6.35}$$

We note that $x(t)$ has the same variation in time that the forcing function does, so that the motion of the mass will always be *in phase* with the action of the excitation. By way of contrast, the speed of the mass is

$$\frac{dx}{dt} = \frac{-\omega F_0}{m(\omega_0^2 - \omega^2)} \sin \omega t \tag{6.36}$$

Since the sine and cosine functions are *out of phase* by 90°, we see that the velocity induced in the mass always lags behind the force by a time equal to $t = \pi/2\omega$.

*This is a further generalization of the construction of Eq. (5.23). It is called the *principle of superposition* and it is a characteristic feature of linear problems.

Another very interesting feature of the solution is that the amplitude $x(t)$ appears to become infinitely large as the forcing frequency ω approaches the natural frequency ω_0. That is, if the driving frequency is equal to the natural frequency, the forcing function produces an (apparently) infinite response in the oscillator. This is the phenomenon of *resonance* and it is familiar to anybody who has ever timed the pushes given to a friend on a swing! We see from Eqs. (6.35) and (6.7) that the amplitude of the response can be written as

$$\text{amplitude of } x(t) = \frac{F_0/k}{1 - \omega^2/\omega_0^2}$$

so that we can illustrate resonance by plotting (see Figure 6.13) the dimensionless amplitude kx/F_0 against the dimensionless frequency ratio ω/ω_0. We have also shown there the effect of damping, which we discuss later, but note for now the infinite peak in the amplitude when $\omega = \omega_0$.

We can also see the manifestation of resonance another way, expressed in terms of quantity called *impedance*. The impedance of a system reflects the resistance of a system responding to a given input.

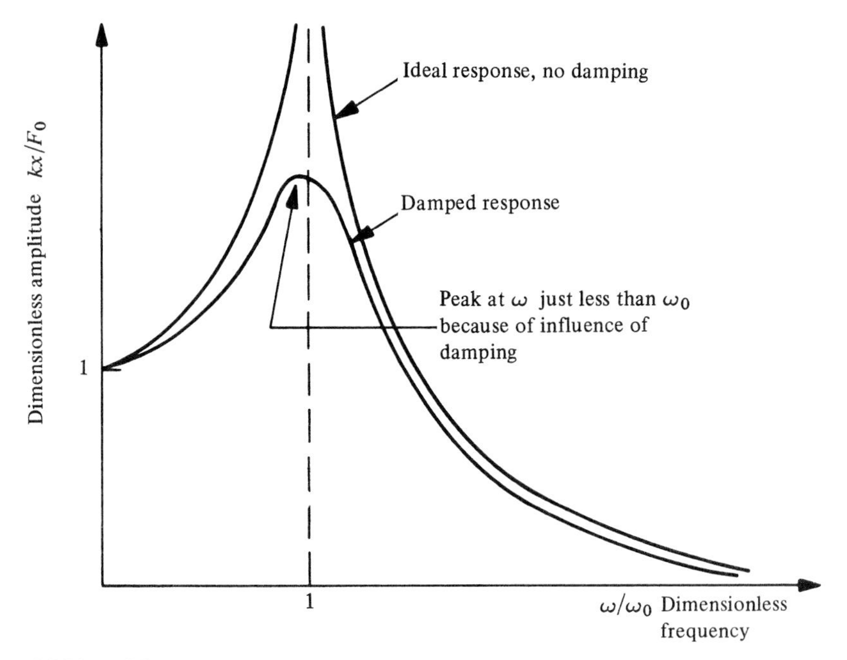

FIGURE 6.13
Response of an oscillator at resonance

For the elementary solution considered here, it is sufficient to take the impedance as the ratio of the magnitude (or amplitude) of the force to the magnitude of the speed.* Using Z to denote the impedance, we find, using Eqs. (6.32) and (6.36),

$$Z = \frac{\text{magnitude of } F(t)}{\text{magnitude of } \dfrac{dx(t)}{dt}}$$

or

$$Z = m\omega_0\left(\frac{\omega_0}{\omega} - \frac{\omega}{\omega_0}\right) \tag{6.37a}$$

and

$$Z = k\left(\frac{1}{\omega} - \frac{m\omega}{k}\right) \tag{6.37b}$$

We have written the formula for the impedance in two different forms in order to demonstrate certain features more easily. Notice first that at resonance, when $\omega = \omega_0$, the impedance vanishes. This confirms the notion of seemingly infinite responses at resonance, since the vanishing of the impedance implies that the speed will be infinitely large even for a small (finite) force. That is, at resonance there is nothing impeding the motion of the mass.

As the second interesting feature we consider two limiting cases of excitation frequency: The first is that of slow excitation, for which $\omega << \omega_0$; the second is that of fast excitation, for which $\omega >> \omega_0$. In the first case, we see from Eqs. (6.37) that the term inversely proportional to ω dominates. Then we can speak of a frequency regime in which the response is *stiffness-controlled*, where

$$Z_{\mathrm{k}} \cong \frac{k}{\omega} \qquad \omega << \omega_0 \tag{6.38}$$

For small values of ω it is not unexpected that the stiffness will control

*The complete definition of impedance can be properly developed if the force–velocity phase shift is accounted for. The standard and most straightforward way to do this is based on the arithmetic of complex numbers, using what physicist Richard Feynman calls "the most remarkable formula in mathematics,"

$$e^{i\omega t} = \cos \omega t + i \sin \omega t \qquad i \equiv \sqrt{-1}$$

Then the phase shift can be included and the impedance can be derived more directly.

the response, for in that regime we are closest to the static limit for which $\omega = 0$. The other regime of interest is that of rapid excitation, for which $\omega >> \omega_0$. Here the second term dominates and the system is said to be *mass-controlled*:

$$Z_{\mathrm{m}} \cong -m\omega \qquad \omega >> \omega_0 \tag{6.39}$$

At high frequencies we would expect the dynamics of a system to be more important, so the fact that the impedance is governed by the mass is reassuring.

At this point it is appropriate to ask the following question: At resonance, when the impedance vanishes and the response seemingly becomes infinite, what actually happens? We know from practical experience—or judgment, or intuition—that the response never actually becomes infinite in magnitude. What we can say in this context is that our model breaks down in one way or another. One possibility is that, as in the case of the simple pendulum, the amplitude becomes so large that the linear model is no longer valid—and the nonlinear model will not predict this type of response! Another possibility is that the model is deficient because we have left out something. In most instances what becomes important at resonance is damping or energy dissipation. The inclusion of the effects of damping requires the kind of complex mathematics referred to in the footnote to page 133, so we will not go into the details here. However, we can cite two important results that are produced as solutions of the equation of motion for a spring-mass-damper subjected to a periodic force:

$$m\frac{d^2x}{dt^2} + R\frac{dx}{dt} + kx = F_0e^{i\omega t} \tag{6.40}$$

The magnitude of the mass motion turns out to be* (again ignoring considerations of phase)

$$|x(t)|^2 = \frac{F_0^2}{m^2(\omega_0^2 - \omega^2)^2 + R^2\omega^2} \tag{6.41}$$

while the magnitude of the impedance is

$$|Z|^2 = m^2\omega_0^2\left(\frac{\omega_0}{\omega} - \frac{\omega}{\omega_0}\right)^2 + R^2 \tag{6.42}$$

We note immediately that in the absence of damping, when $R = 0$, Eqs.

*It would be a good idea for you to check the dimensions in Eqs. (6.41) and (6.42).

(6.41) and (6.42) reduce to their respective counterparts, Eqs. (6.35) and (6.37a). More important, for we ought not to be terribly surprised by this reduction, is that at resonance the impedance does not vanish and the response does not become infinite. That is, when $\omega = \omega_0$

$$|x(t)|^2_{\text{resonance}} = \left(\frac{F_0}{R\omega_0}\right)^2 \tag{6.43}$$

and

$$|Z|^2_{\text{resonance}} = R^2 \tag{6.44}$$

We see clearly from Eq. (6.43) that it is the absence of damping that produces an infinite response in the linear oscillator.* Equation (6.44) shows that the impedance vanishes at resonance only if there is no energy dissipation. Further, this result confirms explicitly our concept of damping as being resistive to motion—or providing an impediment to free motion. Finally, it means we could add to Table 6.1 another line in which we could relate the mechanical impedance (in dimensions of force/velocity) to resistance (in electrical dimensions, ohms; in mechanical dimensions, force/velocity).

SUMMARY

In this chapter we have presented an extended discussion of the simple harmonic oscillator, with and without damping, with and without a forcing function. We have shown many examples of how the oscillator recurs time and again in describing physical problems, including building vibrations, automobile suspensions, the acoustic resonator, the cyclotron, and the classic *RLC* electrical circuit. In discussing the latter, the very useful electrical-mechanical analogy was developed.

We have also discussed the solution to the forced oscillator problem when the forcing function is harmonic. Thus we were able to bring out the concepts of resonance and impedance. In discussing impedance we then identified various regimes of oscillator response which were controlled by the stiffness, damping, or mass of the system.

We have tried to stress throughout the commonality of not only the mathematics, but also of much of the physics. That is, to develop the oscillatory behavior discussed in this chapter a system must have at least one stiffness-controlled element that stores (potential) energy and

*Care is needed in generalizing such a statement. For example, for large swings of a pendulum the nonlinear equations must be used and there the response will remain finite even with $R = 0$.

at least one mass-controlled element that "stores" (kinetic) energy. The stiffness may take many forms, not all immediately obvious, but there must be a device that stores (potential) energy.

Problems

6.1 (a) Find the acoustical impedance Z of a Helmholtz resonator as a function of ρ_0, A, l, ω, and V_0.
(b) What are the dimensions of Z?

6.2 When charged particles are accelerated in a cyclotron, they travel in circles whose radii r depend on their speed v and the magnetic field strength B such that

$$r = \frac{mv}{qB}$$

where m is the mass and q the charge of the particle. At each half cycle the speed, and therefore the energy, is increased so that the particles execute forced simple harmonic oscillation in circles of increasing radii.
(a) At what angular frequency (resonant frequency) ω must the energy be supplied?
(b) What is the impedance of the system?
(c) Show that the rate of change of energy in the system is positive and of the form

$$\frac{dE}{dt} = \frac{q^3B^3(r_2^2 - r_1^2)}{2\pi m^2}$$

6.3 A simple seismograph is shown in the accompanying figure. If y is used to denote the displacement of m relative to the earth, and η

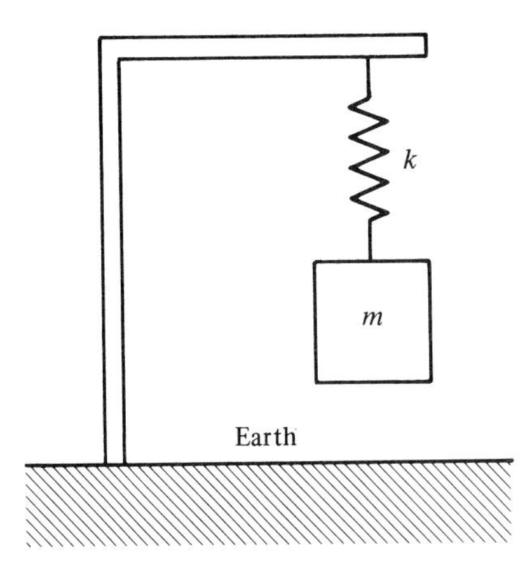

to denote the displacement of the earth's surface itself relative to the fixed stars, the equation of motion of the mass is

$$\frac{d^2y}{dt^2} + \frac{R}{m}\frac{dy}{dt} + \frac{k}{m}y = -\frac{d^2\eta}{dt^2}$$

(a) If $\eta = C \cos \omega t$, write an expression for y (steady-state vibration). Assume C is constant for all ω.
(b) Sketch the amplitude A of the displacement y as a function of ω.

6.4 A long-period seismometer has a mass m of 0.01 kg, a period of about 30 sec, and a damping quality factor $Q = \omega_0 m/R$ of about 3. An earthquake causes the earth's surface to oscillate with a period of 15 min. The maximum acceleration of the earth is 2×10^{-9} m/sec^2. Find the amplitude of the seismometer oscillations.

6.5 Show that Eq. (6.42) is dimensionally correct.

6.6 Consider a forced oscillator with damping where the driving force is $F = F_0 \cos \omega t$ and the displacement $x = A \cos(\omega t - \delta)$. Here, δ is the *phase angle* by which the driving force F leads the resulting displacement x. Find the *average* power input needed to maintain the forced oscillations. (Power equals force times velocity.)

6.7 For the forced oscillator in Problem 6.6, let the phase angle δ be $\pi/2$ rad, let ω_0 be 500 rad/sec, and let a damping constant Q be 4 when (with $\sin \delta = kA\omega/F_0\omega_0 Q$)

$$A = \frac{F_0}{k} \frac{\omega_0/\omega}{\left[\left(\frac{\omega_0}{\omega} - \frac{\omega}{\omega_0}\right)^2 + \frac{1}{Q^2}\right]^{1/2}}$$

(a) Plot the average power input found in Problem 6.6 against the frequency ω of the driving force.
(b) Find the width $\Delta\omega$ of the power resonance curve in part (a) at $\frac{1}{2}\bar{p}_{\max}$. This range of frequencies may be considered as the band of frequencies over which resonance effectively occurs.

6.8 (a) Repeat the calculations in Problem 6.7 for the same oscillator with less damping; that is, with $Q = 6$.
(b) What does a comparison of the two resonance widths ($\Delta\omega$) in Problems 6.7 and 6.8(a) reveal about the effect of damping on resonance?

6.9 List resonant systems that we observe in nature. Cover as wide a range as possible.

6.10 A weight hanging on the end of a spring causes a static deflection $x_{st} = W/k$. If the static deflection is measured in inches, use this static relationship to show that the resonant frequency in hertz (cycles per sec) of a spring–mass oscillator is

$$f = \frac{\omega}{2\pi} = \frac{3.13}{\sqrt{x_{st}}}$$

6.11 A bridge is 100 m long and is supported by steel beams whose modulus of elasticity E is 2 N/m^2 and whose moment of inertia I of the cross-sectional area is 0.002 m^4. If the mass of the bridge is 10^5 kg, what is its natural frequency of oscillation when a weight of $1.8(10^5)$ N causes it to deflect 0.01 m?

6.12 A group of 200 soldiers weighing $1.8(10^5)$ N is marching across the bridge in Problem 6.11. Their right feet touch the bridge every 0.9 sec, forcing the bridge to oscillate. What would an observer see happening in these circumstances? Support your answer.

7

TRAFFIC FLOW MODELS

The last several years have seen the development of very sophisticated mathematical models for problems that originate in fields of human endeavor other than physics. In fact, some of these models make use of analogies with physical problems. One set of models has to do with *traffic flow theory*, an area in which the attempt is being made to analyze and predict the flow of traffic on roads and highways (which are often referred to as arteries, to point out yet another analogy—to the flow of blood). There are two generic types of models that treat the traffic flow problem at different levels.

The first type of model is the *macroscopic* model; it treats traffic flow essentially as the flow of a fluid and deals with gross or averaged variables for a whole line of traffic. Thus an investigator would be concerned with the rate of flow (the number of cars going past a fixed point per unit of time), the speed of the traffic flow (the distance covered per unit time), and the density of traffic (the number of cars in a column of given length). Therefore, in this model structure, we are concerned with the flow of so many particles (cars) that we speak of a *field* or *continuum* of cars moving in a tube or stream called a road or lane of traffic.

This is in constrast to the *microscopic* models of traffic flow where we are concerned with the interactions of individual cars in a line of traffic. Specifically, we will develop here alternative models that describe the acceleration of "follower" cars as functions of the distance between the "leader" and the follower, the relative speed of the two cars, and the reaction time of the driver of the follower car. These models are called *car-following* models and, after a brief overview of the macroscopic theory of traffic flow, we shall investigate several car-following models and we shall then relate these models to the macroscopic variables.

MACROSCOPIC TRAFFIC FLOW THEORY—I. CONTINUUM HYPOTHESIS

As we have mentioned, the macroscopic theory of traffic flow is based on an analogy with the flow of a fluid in a tube. That problem is treated in the science of *hydrodynamics* and the mathematics involved is considerably more complex than what we wish to deal with here. However, some of the basic notions can be sketched out without solving the partial differential equations of fluid mechanics, and in a form directly applicable to the traffic flow problem.

We first need to develop some basic definitions. Consider the line of autos, represented by blocks, displayed in Figure 7.1. We have shown a series of cars moving to the right and we have indicated that each can be identified by a coordinate value x_i measured along an axis parallel to the highway. The position of each car will vary with time, so that $x_i = x_i(t)$; therefore we can calculate the speed dx_i/dt and the acceleration d^2x_i/dt^2 of the ith car at any time t. For the case when there are a great many cars present—or as in a fluid where there are many, many particles present—it is easier to speak of a velocity* field, in which we assign to every point along the x axis a unique velocity $v(x, t)$. Thus we replace the line of discrete cars at the points $x = x_i$ by an infinite sequence of points, each of which now has a unique velocity expressed by the continuous function $v(x, t)$. This first element of the *continuum hypothesis*† also restricts our model so as to eliminate overtaking and passing, since if one auto passes another we would have to have a point on the axis with two different velocities.

It is also useful to characterize the number of cars present for a given analysis. We can do this in two ways. One is to stand at a fixed point on

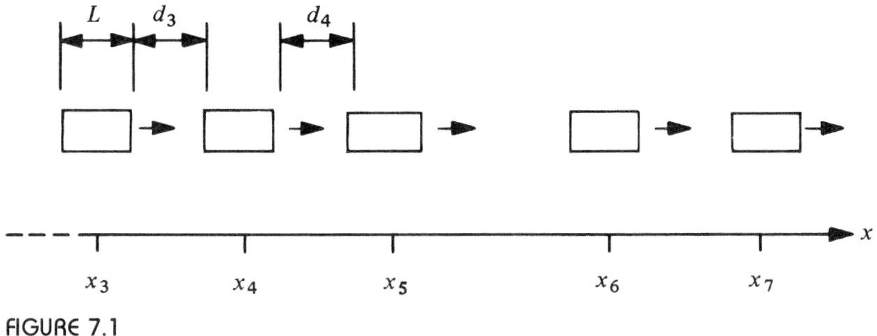

FIGURE 7.1
Coordinates of autos on a highway

*In what follows we shall use speed and velocity interchangeably, even though the latter is generally used to denote vector quantities.

†See the discussion in Chapter 4.

a road and count the number of cars going past in a certain time interval. This would give us a count of the *traffic flow* q expressed in terms of a number of vehicles per unit time. However, we can also count the number of cars in a given length of roadway at any instant and can thus determine the *density* ρ expressed in terms of number of vehicles per unit length. We can do this in a practical sense by taking aerial photographs of a given stretch of road and then counting the number of cars in that stretch. In making both of these observations, however, whether we are counting the traffic flow q or the traffic density ρ, we must ask whether we are counting over an appropriate length of time or road. If the length is too long, we will be averaging over such a long period of time or stretch of road that meaningful fluctuations are canceled out. Thus, if we analyzed our flow count by the day we would miss the peaks generated by morning and evening rush hours. Similarly, if we did our spatial count over the length of the interstate highway system between New York and Boston, for example, we would miss the buildup at various cities and towns along the way as well as the sparse amounts of traffic that might be expected in rural areas.

On the other hand, if the measuring time or distance is too short, then we can experience wild fluctuations that have no relevance. For example, in a short burst of time just before a traffic light changes no traffic may be counted, whereas shortly thereafter, just after a light change, there could be a huge increase in the number of cars. Similarly, if we did the spatial count over very short intervals of highway length, we might pick up fractions of a car in some intervals and many intervals with no cars at all. Thus our density count could be highly discontinuous. A typical plot of the density as a function of the measuring interval is shown in Figure 7.2. Clearly we want to choose the interval in the region where there are enough cars to have a meaningful but relatively constant *local* density. This allows us to obtain continuous functions for the density ρ and the traffic flow q in much the same way in which we replaced the speeds of individual cars with a velocity field.

In the science of fluid mechanics we invoke the continuum hypothesis in the same way, thus dealing not with individual particles of a fluid but with gross or *averaged* variables. In fact, we use mass density, a mass flow rate, and a velocity field in exactly the same way as we have just outlined for traffic flow. Of course what we are doing here is making use (in a heuristic way) of the concepts of *scaling* when we decide how long an interval is required to observe "real" changes in what is being measured.

Although the major focus of this chapter will be on car-following models where, of necessity, we must deal with individual cars, the *continuum hypothesis* used in the development of the velocity field and the density function is a major conceptual device used many times over

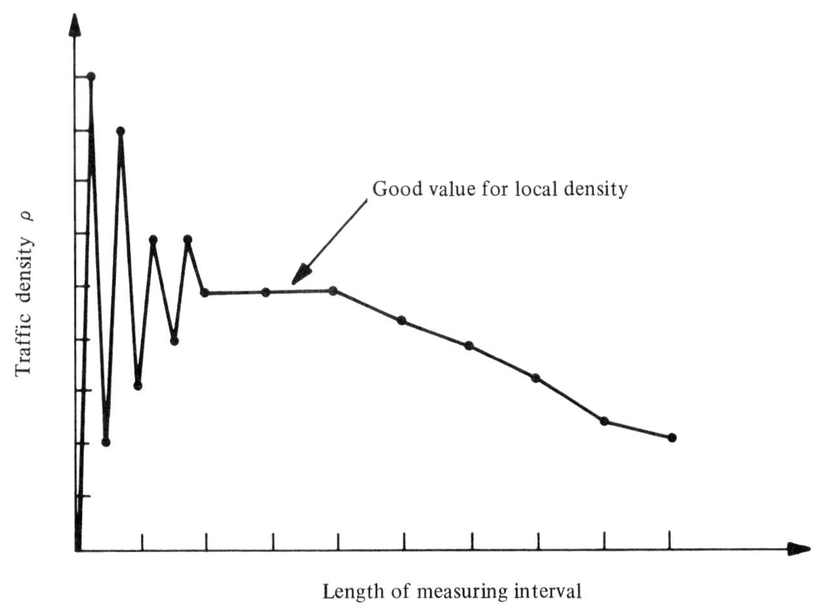

FIGURE 7.2
Typical variation of density with interval length

in the sciences. We will relate the so-called *discrete* model (so named because it deals with individual vehicles) of car-following to the continuum model at the end of this chapter.

MACROSCOPIC TRAFFIC FLOW THEORY—II. THE FUNDAMENTAL DIAGRAM

In this section we shall see how the continuous macroscopic variables—the flow q, the density ρ, and the velocity v—are related, and we shall try to identify the appropriate laws that govern their behavior. In the first instance we note from the dimensions of q, ρ, and v that it would not be dimensionally surprising* if

$$q = \rho v \tag{7.1}$$

However, does this make "physical" sense? It does indeed, and it can

*Remember that q is the number of cars per unit time, ρ is the number of vehicles per unit length, and velocity is length/time.

be demonstrated by a very simple argument. If we counted the number of cars going past a given point on a road during a small interval of time Δt, we could calculate the total number of cars passing using two different measures.

The first measure is straightforward enough: If the traffic flow at the measuring point is q vehicles per unit time, then in an interval Δt the number of vehicles passing is $q\ \Delta t$. For the second measure we note that in an interval Δt a vehicle having a speed of v will cover a distance equal to $v\ \Delta t$. Then the number of vehicles passing through that distance is just the product of the density ρ times the distance or $\rho v\ \Delta t$. Hence, because these two measures must give the same result, we find upon equating them that we can confirm the validity of Eq. (7.1). In general, of course, all the variables ρ, v, and q are functions of the roadway coordinate x and the time t. Because of this relation the derivation of Eq. (7.1) that we have just given is somewhat heuristic, since we have (implicitly) treated $v(x, t)$, for example, as an averaged constant in computing the distance traveled during the time interval Δt. Nevertheless, in spite of having ignored some rigorous fine points, the general relation between flow, density, and velocity is

$$q(x, t) = \rho(x, t)v(x, t) \tag{7.2}$$

By virtue of this result we can identify as the two fundamental traffic variables the density $\rho(x, t)$ and the velocity $v(x, t)$, since the flow $q(x, t)$ can be obtained from Eq. (7.2). However, in order to develop a complete model—perhaps for the purpose of predicting the velocity and the density downstream from some point on a road where these quantities are known—we need some more relationships among these variables. These further relationships come from the "principle of *conservation of cars*" and from assumed relationships between the velocity and the density. The conservation law is directly analogous to the principle of mass conservation or the statement that "what goes in must come out or stay." The velocity–density relations are the results of attempts to describe or observe how fast people drive their cars in varying traffic conditions. We shall return to this point when we discuss discrete car-following models.

The law of conservation of cars is easily stated. If we consider two points on a road denoted by $x = a$ and $x = b$, respectively, where $a < b$, then (assuming that no cars are destroyed or created within this interval) changes in the number of cars in the interval $a \leq x \leq b$ can be produced only by crossings into and out of the interval at the end points. The number of cars entering per unit time at $x = a$ is simply the flow $q(a, t)$, while the number of cars leaving the interval at $x = b$ is $q(b, t)$. Hence, if N is the total number of cars in the space $a \leq x \leq b$, the rate of change of N with time must be

$$\frac{dN}{dt} = q(a, t) - q(b, t) \tag{7.3}$$

On the other hand, we know that the total number of cars in that interval is simply the integral of the density $\rho(x, t)$ over the length of the interval:

$$N(t) = \int_{x=a}^{x=b} \rho(x, t)\, dx \tag{7.4}$$

Thus, the conservation of cars requires that

$$\frac{d}{dt} \int_a^b \rho(x, t)\, dx = q(a, t) - q(b, t) = \int_{x=b}^{x=a} \frac{\partial q(x, t)}{\partial x}\, dx \tag{7.5}$$

This conservation law can also be shown to be precisely equivalent to the following *partial differential equation:*

$$\frac{\partial \rho(x, t)}{\partial t} + \frac{\partial q(x, t)}{\partial x} = 0 \tag{7.6}$$

This result, which relates $\rho(x, t)$ to $q(x, t)$ and so fills in one more step in the macroscopic traffic model, is exactly the same as the *continuity equation* of fluid mechanics. That equation is a statement of the conservation of mass—assuming again that mass is neither created nor destroyed in whatever process may be under consideration.

We now have in Eqs. (7.2) and (7.6) two of the three equations needed to relate our three flow variables. By substituting Eq. (7.2) into Eq. (7.6) we can reduce our problem to having only two unknowns with one equation; that is,

$$\frac{\partial \rho}{\partial t} + \frac{\partial}{\partial x} (\rho v) = 0 \tag{7.7}$$

We must now somehow relate the car velocities to the density of vehicles on the road, so that we can think about solving Eq. (7.7). However, we have to reckon with the fact that we are no longer looking for a mechanical law because we know that the vehicle velocities depend on their drivers! In other words, we are now trying to model human judgment. How does the individual driver respond to the traffic conditions around him/her? It is clearly a case of responding to stimuli, the question being: To what stimuli? Clearly people respond to what the vehicle in front of them is doing, although a macroscopic model cannot account for this directly. Drivers also respond to the density of traffic around them, speeding up in light traffic and slowing down (perhaps involuntarily) in heavy traffic. Thus, we are tempted to postulate that the velocity of a vehicle depends only on the density of traffic, that is,

$$v \equiv v(\rho) \tag{7.8}$$

If there are no—or very few—other cars on the road, we would expect the driver to maintain the greatest speed, $v = v_{max}$. Further, as the density increases, the velocity is presumed to decrease. Finally, at some maximum density (bumper-to-bumper traffic) traffic crawls to a halt, so that $v = 0$ at $\rho = \rho_{max}$. We can summarize these assumptions as mathematical statements

$$v(\rho = 0) = v_{max} \tag{7.9a}$$

$$\frac{dv}{d\rho} \leq 0 \tag{7.9b}$$

$$v(\rho = \rho_{max}) = 0 \tag{7.9c}$$

or in the form of a schematic, as in Figure 7.3. Note that the precise shape of the curve $v = v(\rho)$ is not specified; only the end points and the sign of the slope are known.

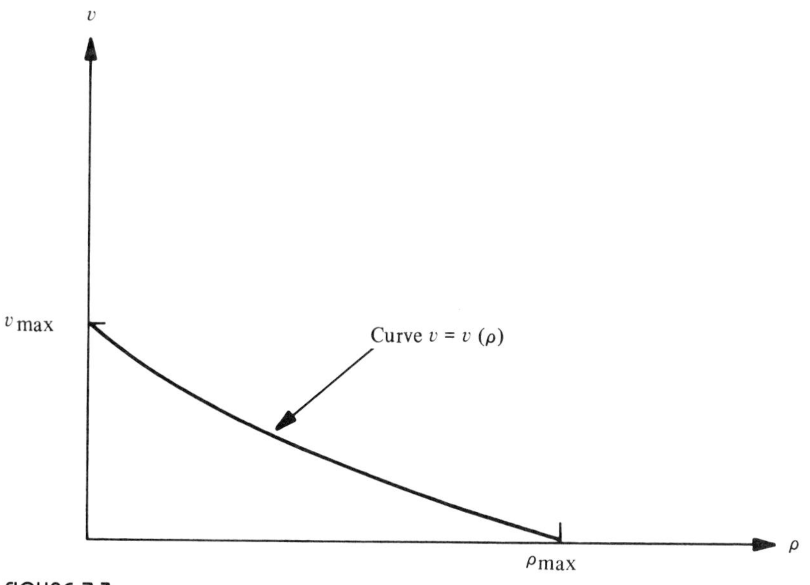

FIGURE 7.3
Schematic of velocity-density curve

Although the assumption embodied in Eqs. (7.8) and (7.9) does not account for all possibilities, it has proven to have a fair range of validity while traffic is accelerating or decelerating. Other factors that might be considered, in addition to those just mentioned, are related to the delay or reaction time between the driver's perception of a given density and his or her consequent action, and to the fact that a driver responds to changes in the traffic density several cars ahead rather than the density immediately ahead. However, notwithstanding these observations, we can still use the model for the purpose of analyzing traffic flow.

From the viewpoint of the traffic engineer who wishes to design a road and its appurtenances (including entrance and exit ramps, signal systems, tollbooths, etc.), the principal variable is the traffic flow q, which indicates the capacity of the system to be designed. If the velocity is taken to be *homogeneous*, that is, independent of the time or position on the road, we can calculate the flow by combining Eqs. (7.1) and (7.8):

$$q = \rho v(\rho) \tag{7.10}$$

Therefore, for this macroscopic model, the traffic flow ultimately depends only on the traffic density ρ.

We know, as with the velocity-density relation, some of the characteristics of the flow-density relation. For example, even though at $\rho = 0$ we have $v = v_{\max}$, it is obvious from Eq. (7.10) that $q(\rho = 0) = 0$. Also, at the maximum density ($\rho = \rho_{\max}$) we know that traffic halts, so that $v(\rho = \rho_{\max}) = 0$ and thus $q(\rho_{\max}) = 0$. Also, the traffic flow must always be positive for $0 < \rho < \rho_{\max}$. Therefore, the general relationship between flow and density must be in the shape of the *fundamental diagram of road traffic* that is given in Figure 7.4. The maximum flow $q_{\max}$ is termed by traffic engineers the *capacity* of the road.

To make some of these ideas a bit more concrete, let us see what happens if we assume a *linear* velocity-density relation, that is, if we let

$$v = v_{\max}\left(1 - \frac{\rho}{\rho_{\max}}\right) \tag{7.11}$$

This assumption clearly satisfies all the conditions stated in Eqs. (7.9). In addition it gives a *parabolic* relation between flow and density (from Eqs. (7.11) and (7.10)):

$$q = v_{\max}\left(\rho - \frac{\rho^2}{\rho_{\max}}\right) \tag{7.12}$$

The maximum flow rate is then found as the point on the fundamental

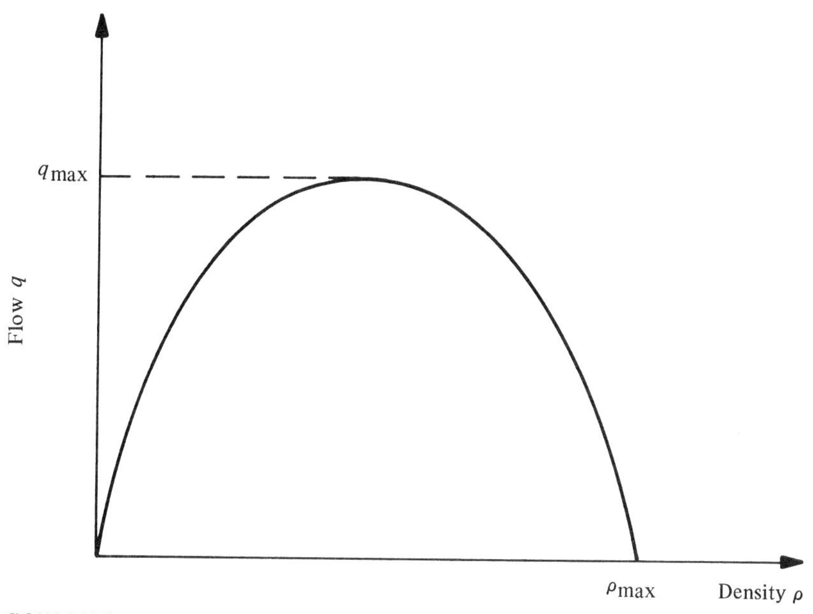

FIGURE 7.4
Fundamental diagram of road traffic

diagram where the slope $dq/d\rho$ vanishes. Then,

$$\frac{dq}{d\rho} = v_{max}\left(1 - \frac{2\rho}{\rho_{max}}\right)$$

so that the value of the density at which the flow peaks is $\rho = \rho_{max}/2$ and the maximum flow is

$$q_{max} = \tfrac{1}{4} v_{max} \rho_{max} \tag{7.13}$$

Finally we can ask whether the linear relation given by Eq. (7.11) has any validity beyond its providing a nice example for demonstration purposes. In fact, the linear velocity–density relation has been shown to be good in studies conducted in several tunnels in New York City (viz., the Lincoln, Holland, and Queens–Midtown Tunnels), and it provides a good approximation for the central portion of an empirical velocity–density curve such as the one displayed in Figure 7.5.

LINEAR CAR-FOLLOWING MODELS

We now turn from the macroscopic problem, where we have dealt with averaged variables, to the microscopic problem, where we will look at

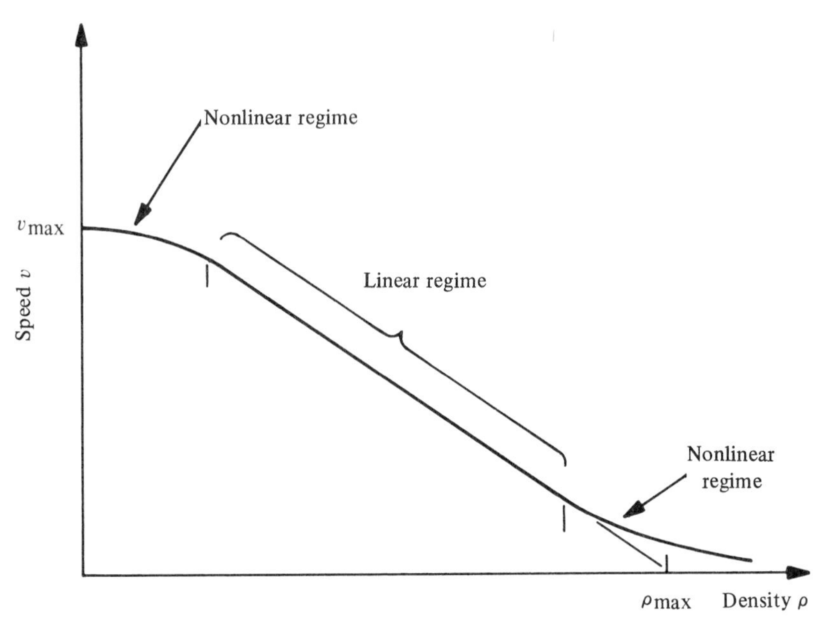

FIGURE 7.5
Linear and nonlinear regimes of velocity-density curve

individual cars. Our aim is to see whether we can use a discrete car-following model to generate useful and applicable velocity-density curves. We also note that the problem here is one of modeling psychological behavior rather than mechanical performance. This is because the models must describe—and even predict—the response of drivers to the conditions around them, so the models must predict the *response to stimulus*. The stimulus could be the relative distance between a following car and the car in front, the relative velocity of the two cars, or the relative acceleration. The response will depend in part on the *sensitivity* of the responder to a given stimulus. Also, it may often be true that the response is not immediate, so that a time delay must be incorporated into the model.

The basic equation of car-following for one-lane traffic with reasonably high density and no passing or overtaking is the psychological one:

$$\text{response} = \text{sensitivity} \cdot \text{stimulus} \tag{7.14}$$

The response will be taken as the acceleration of the following vehicle, denoted here by d^2x_F/dt^2, while the stimulus will be taken as the difference between the velocity of the lead vehicle dx_L/dt and the velocity of the trailing vehicle dx_F/dt. For now, the sensitivity will be incorporated in the symbol λ and we shall defer until later discussion of how λ might be modeled or evaluated. Then the response is

$$\frac{d^2x_F}{dt^2} = -\lambda\left(\frac{dx_F}{dt} - \frac{dx_L}{dt}\right) \tag{7.15}$$

Here, with $\lambda > 0$, if the following car is going faster than the lead car, the following car will decelerate to avoid hitting the car in front—or at least so we hope! Note that λ has dimensions of $(\text{time})^{-1}$. If we wanted to build a reaction time into this model we could write Eq. (7.15) as

$$\frac{d^2x_F(t+T)}{dt^2} = -\lambda\left(\frac{dx_F(t)}{dt} - \frac{dx_L(t)}{dt}\right) \tag{7.16}$$

Here T is the time it takes the driver of the following car to respond to the situation; that is, T can be taken to be the driver's *reaction time*.

For the case where λ is a constant we can easily integrate Eq. (7.16) to obtain the equation for a *linear* car-following model, a result that can be verified by differentiation:

$$\frac{dx_F(t+T)}{dt} = -\lambda[x_F(t) - x_L(t)] + D_F \tag{7.17}$$

and where D_F is an arbitrary constant of integration with the dimensions of speed. This result relates the velocity of the following car to the spacing between cars. This result is also one that we can convert into a velocity-density result of the type in Eq. (7.8). We first assume that the traffic flow is *steady state*—all cars are equidistant from one another and so all are traveling at the same speed. Thus

$$\frac{dx_F(t+T)}{dt} = \frac{dx_F(t)}{dt}$$

and it is also true that $D_F = D$ is the same for all follower-leader pairs, that is, for all cars on the road. We also note that the density ρ can be defined as

$$\rho = \frac{1}{\text{spacing between cars} + \text{one car length}}$$

so that

$$x_L(t) - x_F(t) = \frac{1}{\rho} \tag{7.18}$$

Since by definition $v = dx_F/dt$, it follows from Eqs. (7.17) and (7.18) that for steady-state conditions

$$v = \frac{\lambda}{\rho} + D \tag{7.19}$$

The constant D can be determined from the condition that the velocity is zero when the density is a maximum [Eq. (7.9c)]. Hence

$$0 = \frac{\lambda}{\rho_{\max}} + D$$

from which it follows that

$$v = \lambda\left(\frac{1}{\rho} - \frac{1}{\rho_{\max}}\right) \tag{7.20}$$

Equation (7.20) is sketched in Figure 7.6.

The result sketched in Figure 7.6 is not bad for values of ρ sufficiently greater than zero. In such a regime the last two of the conditions in Eqs. (7.9) are satisfied. Clearly the result of an infinite velocity at zero densities is not credible, so an adjustment will have to be made for small values of ρ. The easiest way to do this is to require that for all values of ρ less than some critical value ρ_c the velocity should equal the maximum velocity $v_{\max}$. This is illustrated in Figure 7.7 and by Eqs. (7.21):

$$v \begin{cases} = v_{\max} & \rho < \rho_c \\ = \lambda\left(\dfrac{1}{\rho} - \dfrac{1}{\rho_{\max}}\right) & \rho > \rho_c \end{cases} \tag{7.21a}$$

where

$$\rho_c = \left(\frac{v_{\max}}{\lambda} + \frac{1}{\rho_{\max}}\right)^{-1} \tag{7.21b}$$

Note that the dimension of $v_{\max}/\lambda$ is velocity/(time)$^{-1}$ = length, which is the dimension of $1/\rho_{\max}$, so Eq. (7.21b) is dimensionally correct. The corresponding flows are, in the two density regimes,

$$q \begin{cases} = \rho v_{\max} & \rho < \rho_c \\ = \lambda\left(1 - \dfrac{\rho}{\rho_{\max}}\right) & \rho > \rho_c \end{cases} \tag{7.22}$$

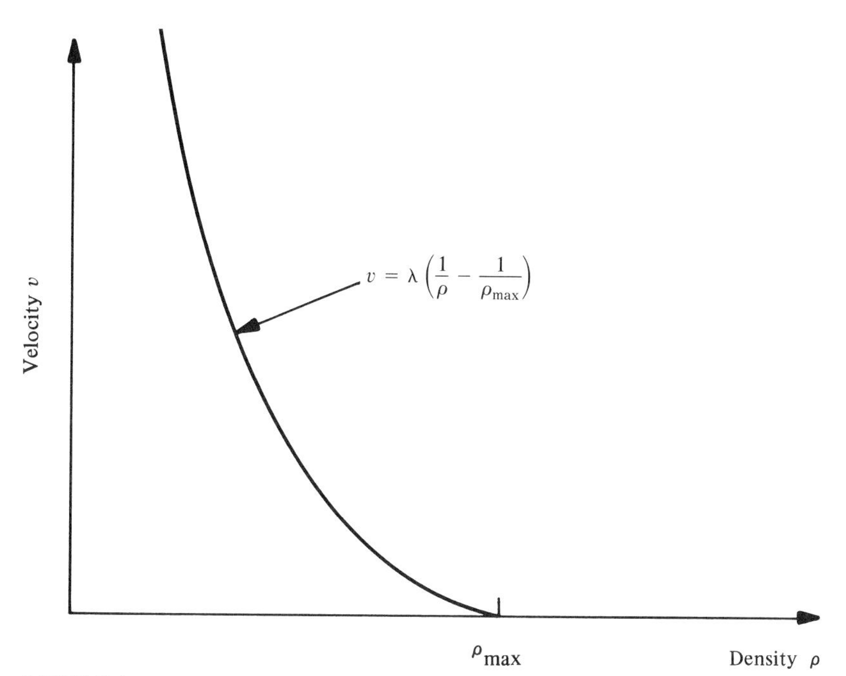

FIGURE 7.6
Sketch of velocity-density distribution of Eq. (7.20)

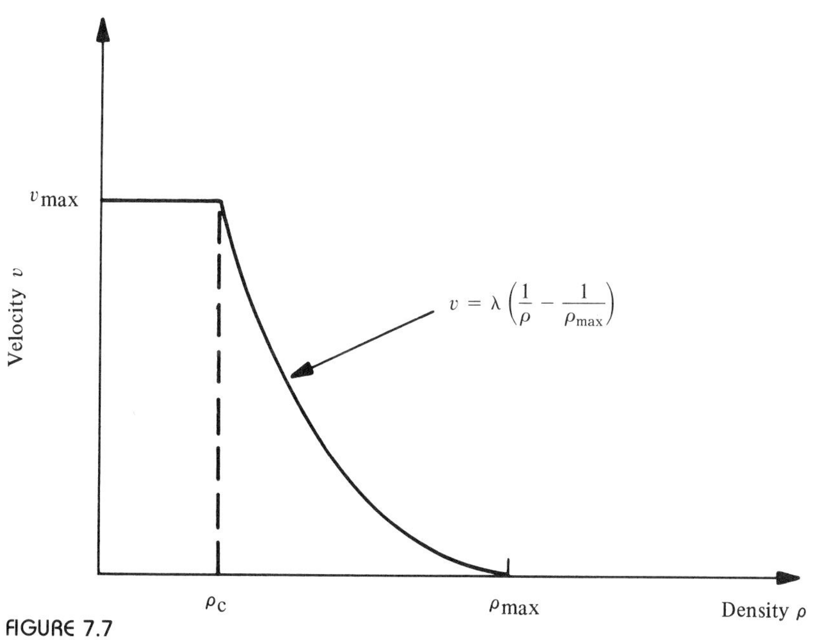

FIGURE 7.7
Modified velocity-density relationship

Unfortunately, although this linear model is straightforward and easy to manipulate analytically, it does not stand up well to observations that have been made in many practical situations. Therefore, in a later section we shall see how the model in Eq. (7.16) can be generalized by lifting the restriction that λ = constant. Before proceeding to these generalizations, however, we shall demonstrate one more phenomenon with the linear model: a *stability analysis*.

STABILITY ANALYSIS FOR LINEAR CAR-FOLLOWING MODELS

A stability analysis is desirable, even necessary, because it allows us to investigate further how the cars interact beyond the capacity calculations just outlined. Thus, if the lead car does not move at a constant speed but oscillates instead about some mean value of the speed, what are the implications for the follower car? Can it keep a safe distance behind? Will it perhaps hit the leader car? Let us see. We note, though, that in order to complete the details of this analysis properly we would have to use the arithmetic of complex numbers. However, we shall—since we know the answer—proceed to the result with a little bit of fudging.

First we shall rewrite Eq. (7.16) in terms of the velocities of the following car ($v_F = dx_F/dt$) and the leading car ($v_L = dx_L/dt$). Then

$$\frac{dv_F(t + T)}{dt} = -\lambda[v_F(t) - v_L(t)] \tag{7.23}$$

Now we assume that the leading car oscillates with amplitude v and frequency ω about a constant speed v_0. Thus we are assuming that

$$v_L(t) = v_0 + v \cos \omega t \tag{7.24}$$

Note that as long as $v < v_0$, the lead car never stops. For the following car we assume a similar form except that we include an *amplification factor* ϵ on the magnitude of the oscillation and a *phase shift* δ in the oscillatory argument:*

$$v_F(t) = v_0 + \epsilon v \cos(\omega t + \delta) \tag{7.25}$$

*Note the difference, particularly the phase shift δ, between Eq. (7.25) and the simple harmonic oscillator solution (6.34), in which no phase shift is required. This difference is due to the different forms of the left-hand sides of the governing equations; that is, Eqs. (6.32) and (7.23). Equation (7.23) is a first-order differential equation that contains a *time delay* term on the left-hand side.

Equation (7.25) can be shown by substitution to be a solution to Eqs. (7.23) and (7.24) if

$$\epsilon\omega \sin(\omega t + \delta + \omega T) = \lambda[\epsilon \cos(\omega t + \delta) - \cos \omega t]$$

Or, expanding the sine of the sum of two angles, noting that $\cos \omega t = \cos(\omega t + \delta - \delta)$, transposing terms, and letting $\phi \equiv \omega t + \delta$, we have

$$\begin{aligned}(\epsilon\omega \cos \omega T + \lambda \sin \delta) \sin \phi \\ + (\epsilon\omega \sin \omega T - \lambda\epsilon + \lambda \cos \delta) \cos \phi = 0\end{aligned} \tag{7.26}$$

Now Eq. (7.26) can have a solution only if the two terms in parentheses both vanish independently.* Thus we obtain two conditions to determine the two unknowns, the amplification ϵ and the phase shift δ. After some further manipulation (noting in particular that we can solve for $\sin \delta$ and $\cos \delta$ and that $\sin^2 \delta + \cos^2 \delta = 1$) we can solve for the amplification factor ϵ and find it in the form

$$\epsilon = \left[\left(\frac{\omega}{\lambda}\right)^2 - 2\frac{\omega}{\lambda} \sin \omega T + 1\right]^{-1/2} \tag{7.27}$$

For the following car to exhibit stable behavior we would have to have $\epsilon < 1$, so that the amplitude of the oscillations of the follower car is smaller than the corresponding amplitude of the leading car. This in turn implies that the term in parentheses in Eq. (7.27) must be greater than unity, so that

$$\left(\frac{\omega}{\lambda}\right)^2 - 2\frac{\omega}{\lambda} \sin \omega T > 0$$

or

$$\frac{\sin \omega T}{\omega T} < \frac{1}{2\lambda T} \tag{7.28}$$

Since $\sin x/x$ is always less than unity (Chapter 4), it follows that the inequality (7.28) is satisfied for any frequency of oscillation ω if the sensitivity λ and the reaction time T are such that

$$\lambda T < \tfrac{1}{2} \tag{7.29}$$

Thus, if the reaction time T is less than the reciprocal of twice the sensitivity—and recall that the dimensions of sensitivity are $(\text{time})^{-1}$,

*This is basically because $\sin \phi$ and $\cos \phi$ vanish for different values of ϕ.

so that the result (7.29) is dimensionally correct—then the following driver is able to maintain control of his vehicle in a stable fashion. This result was obtained, we emphasize, using a linear model where the principal *control variable* is the relative velocity of the leading and following cars. Experiments conducted at the General Motors Research Laboratories indicate that this model has some validity and that some typical numercial values for the parameters involved show a reaction time $T = 1.5$ sec and a sensitivity $\lambda = 0.37\ \text{sec}^{-1}$. In the case of nonlinear models the situation—and the mathematics—is much more complicated.

We should also note that the foregoing investigation has dealt only with *local stability*, where we are concerned with just two cars, a leader and a follower. The analysis can be extended to an evaluation of the stability of a whole line of cars.

NONLINEAR CAR-FOLLOWING MODELS

To round out our discussion of traffic flow models, we wish to extend the discrete car-following model to instances where the sensitivity λ is not constant. We shall present a specific model that is based on the proposition that the driver of the following cars responds to the distance between his/her car and the one in front as well as to the relative velocities. Then the sensitivity can be taken to be

$$\lambda = \frac{\lambda'}{x_L(t) - x_F(t)} \tag{7.30}$$

Here λ' is a modified constant of proportionality with units of (length time)$^{-1}$. Then the modified car-following model becomes [see Eq. (7.16)]

$$\frac{d^2x_F(t+T)}{dt^2} = \frac{\lambda'(dx_F/dt - dx_L/dt)}{(x_F - x_L)} \tag{7.31}$$

which is clearly a nonlinear differential equation. However, this equation can be integrated to yield [compare with Eq. (7.17)]

$$\frac{dx_F(t+T)}{dt} = \lambda' \ln|x_F(t) - x_L(t)| + D_F \tag{7.32}$$

For steady-state movement $D_F = D$. Noting the definitions of the density [Eq. (7.18)] and of the velocity of the follower, we see that Eq.

(7.32) can be written as*

$$v = -\lambda' \ln \rho + D \tag{7.33}$$

Again we evaluate the constant D by stipulating that the velocity vanishes when the density is at its maximum. So we finally obtain the velocity–density relation in the form

$$v = -\lambda' \ln \frac{\rho}{\rho_{\max}} \tag{7.34}$$

The corresponding flow is found by applying Eq. (7.10):

$$q = -\lambda' \rho \ln \frac{\rho}{\rho_{\max}} \tag{7.35}$$

The flow reaches its maximum at that point on the flow–density curve where the slope $dq/d\rho$ vanishes. Here for maximum flow we have the following magnitude of velocity [see also Eqs. (7.12) and (7.13)]:

$$v\Big|_{\substack{\text{maximum}\\ \text{flow}}} = \frac{\lambda' e}{\rho_{\max}} \tag{7.36}$$

where $e = 2.71828$.

The results we have obtained here for this nonlinear model are shown only to emphasize that the microscopic car-following models can be generalized and that even the nonlinear models can be usefully developed for applications. In fact a more general model, in which the sensitivity is of the form

$$\lambda_{\text{nonlinear}} = \lambda' \frac{\left[\dfrac{dx_{\text{F}}(t+T)}{dt}\right]^m}{[x_{\text{F}}(t) - x_{\text{L}}(t)]^l} \tag{7.37}$$

has been dealt with in the literature of traffic flow theory for a variety of (integral) values of the exponents l and m. In many of these cases, incidentally, the connection to the macroscopic flow theory can be made and realistic velocity–density curves can be obtained.

We should also note that there is one flaw in the nonlinear model result of Eq. (7.34); that is, it behaves well and satisfies conditions (7.9b) and (7.9c), but as in the linear model the velocity becomes in-

*We have also noted that $\ln(1/\rho) = -\ln \rho$.

finite as ρ tends to zero. This is the principal reason why models typified by Eq. (7.37) are continually being examined and refined.

SUMMARY

In this chapter we have presented both macroscopic and microscopic models of traffic flow. In the former we used average or gross variables, such as traffic flow and density, to characterize the flow of a column of traffic. We introduced scaling and the continuum hypothesis in this analysis as aides in deciding on appropriate lengths of measuring intervals, whether in time or space. The continuum hypothesis also serves as an introduction to the notion of continuous functions, which is a crucial element in the calculus.

We then treated both linear and nonlinear car-following models in which "microscopic" car-to-car interaction was the focus of interest. These models incorporate different models of psychological response to the varying stimuli afforded by the behavior of the leader car. Such a linear model was also examined for stability to see if a following car could be controlled so that its oscillations about a mean speed would not exceed those of the leading car. Finally, we related the microscopic models to the macroscopic to indicate that car-following models could be used to develop velocity-density curves and so could be used to predict highway capacity.

Problems

7.1 Consider the situation where the traffic flow increases as x increases ($\partial q/\partial x > 0$). What does Eq. (7.6) require of the density? Give a physical explanation for this.

7.2 Assuming the average length of a car to be about 5 m, calculate the density of traffic when cars are traveling with two car lengths between them. If traffic is moving at 80 km/hr (50 mph), what is the traffic flow?

7.3 Apply the continuum hypothesis to find a good value for local density at point A for the example depicted here. Consider a road segment 0.5 of a mile long that is divided into 10 smaller, evenly spaced intervals. A graphic illustration of the locations of 40 cars on the 0.5-mile segment (at some instant in time) appears in the accompanying diagram (p. 157), with traffic density represented by the density of the dots. If the cars are of zero length (for simplicity), find the "good" value for local density in terms of the number of cars per mile. (See also Chapter 4.)

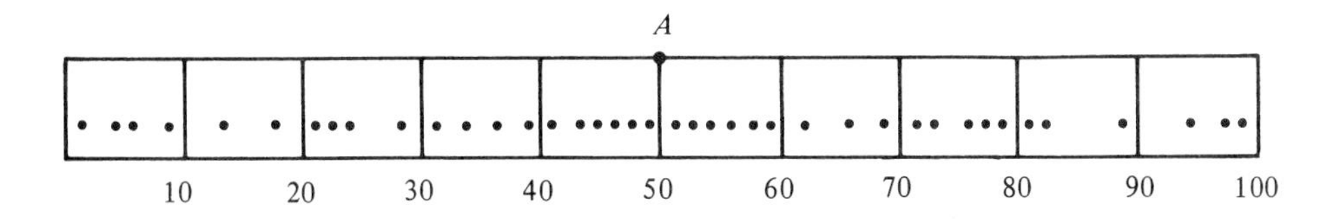

7.4 (a) Assume the velocity is linearly dependent on the density such that $v(\rho) = a + b\rho$. Using the assumptions in Eqs. (7.9), find a and b in terms of the maximum values of the velocity and density.
(b) What is the flow as a function of density?

7.5 Sketch the fundamental diagram of road traffic for the situation in Problem 7.4 where $a = 80$ km/hr and $b = -10^5$ m^2/car · hr.
(a) What are the values of the density and velocity when the flow is maximum?
(b) What is the capacity of the road?

7.6 Consider a flow-density relationship of the form $q(\rho) = \rho(\alpha - \beta\rho)$. Traffic data were used with this relationship to show that the best fit (least squares) occurred when $\alpha = 91.33$ km/hr and $\beta = 1.4$ km^2/car · hr.
(a) What is the maximum density?
(b) What is the maximum velocity?
(c) What is the capacity of the road?
(d) What type of road do you think it is? Explain.

7.7 One car is following another along a highway. The lead car is traveling at 88.5 km/hr (55 mph) and the follower car at 96.6 km/hr (60 mph). For a sensitivity of 0.33 sec^{-1}, what will be the response of the follower car? Is the following driver able to maintain control of his vehicle in a stable fashion if the response time T is 1.8 sec? If $T = 1.5$ sec?

7.8 The data given in the accompanying table were obtained by re-

Velocity (mph)	Density (cars/mile)	Velocity (mph)	Density (cars/mile)
42	44	18	90
40	49	17	95
37	53	16	101
35	58	15	106
32	64	14	112
28	67	13	120
26	69	12	128
23	74	11	139
20	80	10	151
19	85	9	166

cording traffic parameters along a busy stretch of roadway. Sketch the fundamental diagram of road traffic. What is the maximum traffic flow? For what density and velocity does the maximum traffic flow occur?

7.9 For the data in Problem 7.8, plot velocity versus density. Draw an approximate curve and estimate the maximum velocity and the maximum density for this road.

7.10 Using macroscopic traffic flow theory and assuming a linear velocity–density relation, find the speed of traffic when there are three car lengths between lead and follower cars. For this road, $v_{max} = 88.5$ km/hr (55 mph) and $\rho_{max} = 0.22\ m^{-1}$. (Assume a car to be 5 m long.) What is the capacity of the road?

7.11 For a velocity–density relationship similar to that shown in Figure 7.7 and a sensitivity $\lambda = 0.37\ sec^{-1}$, what is the maximum density the road can carry if the density of 25 cars/mile moves at 45 mph?

7.12 If the speed limit is 55 mph in Problem 7.11, what is the critical density? What is the traffic flow of the road (in cars per hour) at the critical density?

8

EXPONENTIAL MODELS

This chapter is devoted to a discussion of *exponential models* and to their application in various fields, including physics, finance, and population studies. The basic characteristic of an exponential model is that it assumes that the rate of change of a parameter, negative (decay) or positive (growth), is directly proportional to the instantaneous value of the parameter. Thus, in addition to applications in the physical sciences, we ought not to be surprised to see applications to problems involving interest rates or population predictions.

EXPONENTIAL BEHAVIOR

As we have indicated, the primary characteristic of exponential growth or decay of a population of objects is the dependence of the rate of change of the population on the instantaneous size of the population itself. As we shall see shortly, this characteristic also implies smooth growth curves that tend to shoot up to very large values in relatively short periods of time. In fact, with the possible exception of radioactive decay problems, we are much more accustomed to seeing exponential *growth* (rather than *decay*) in our everyday affairs. One of the most widely publicized such situations in recent years has been the growth of the world's population and the projections of what the world situation will be like if the rate of growth is not reduced.

We illustrate this situation in the two population projections shown in Figures 8.1 and 8.2. They both exhibit rapid increases in population and, depending on our viewpoint or scale, the increases can be said to occur in a very short period of time. The first curve is based in part on historical data, for the period prior to 1960, and then is continued from

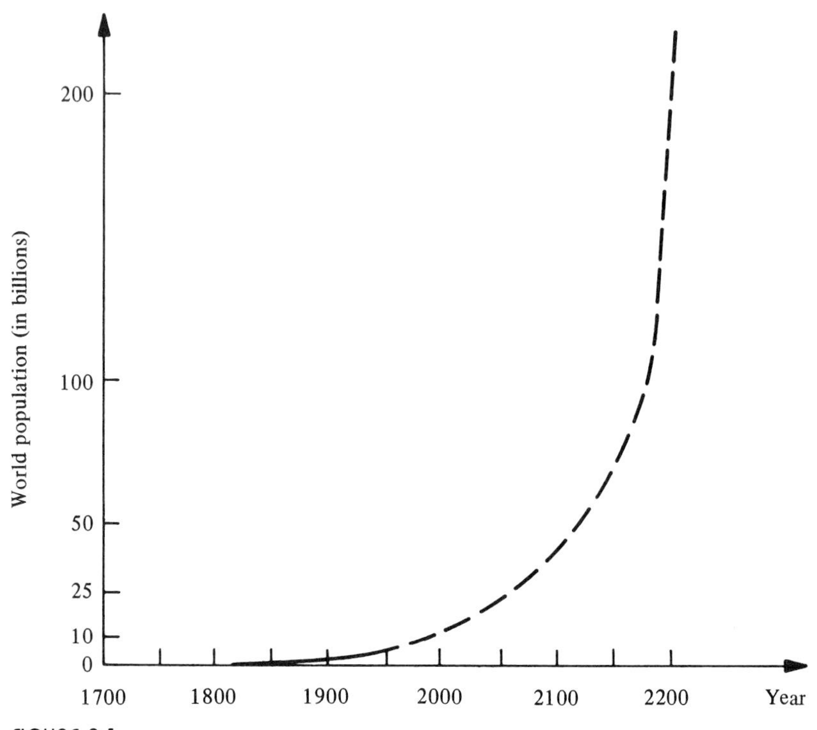

FIGURE 8.1
Historical view (1700–1960) and projection (1960–2165) of world population

the 1960 world population of 3 billion people at an annual growth rate of 2% per year. Prior to the year 1700 the world population was quite small, and it has been growing rapidly since the end of the nineteenth century. We would also be inclined to think that the growth shown in Figure 8.1 is absurd, even though it is based on a very modest growth rate of 2% per year.

If we were to extend the growth of population from Figure 8.1 to the next 700 or 800 years, we would obtain the curve displayed in Figure 8.2. Again, the growth rate is taken to be 2% per year and we are still measuring the population in billions. But we have now expanded the time scale by a factor of two, and look at the numbers on the ordinate scale! These population estimates seem (and almost surely are) totally unrealistic. However, the curve dramatically illustrates the phenomenon of exponential growth: The bigger it is, the faster it grows.

We also emphasize that scaling is very important in looking at exponential growth. One point about the scale has to do with the magnitude of the numbers themselves. For example, a world population of 3 billion in 1960, at a 2% growth rate, becomes a population of 5,630,000

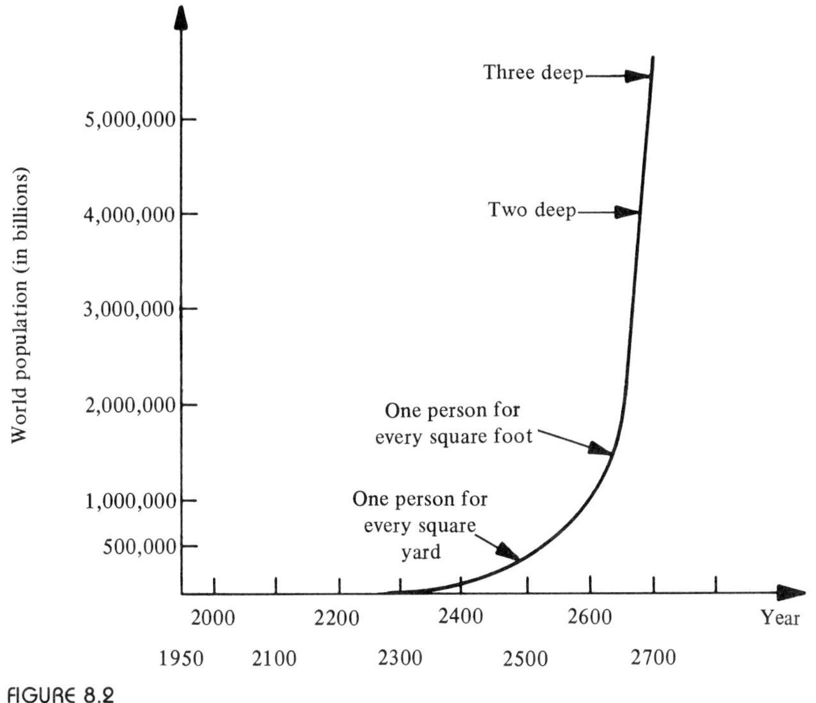

FIGURE 8.2
Projection of world population to 2700

billion in 2692. What does it mean to have

5,630,000 billion people

or

5,630,000,000,000,000 people

or

5.63×10^{15} people

on earth in the year 2692? Is there room for all these people? The total surface area of the earth is 5.028×10^8 km² (5.586×10^{15} ft²), of which only 28% is land. Thus, there is "standing area" of only 1.408×10^8 km² (1.564×10^{15} ft²), so that if each person is allotted 1 ft², people would have to be stacked more than three deep in order to be accommodated. Further, could we even count that many people in a census?

Suppose we could count at a rate of 1000 people per second. Then it would take

$$\frac{5.63 \times 10^{15} \text{ people}}{1000 \text{ people/sec}} = 5.63 \times 10^{12} \text{ sec}$$

to count all those people. This seems like a lot of seconds, and it is:

$$\begin{aligned}\text{counting time} &= 5.63 \times 10^{12}\ \text{sec} \times \frac{1\ \text{min}}{60\ \text{sec}} \\ &\quad \times \frac{1\ \text{hr}}{60\ \text{min}} \times \frac{1\ \text{day}}{24\ \text{hr}} \times \frac{1\ \text{yr}}{365\ \text{days}} \\ &\cong 1.78 \times 10^{5}\ \text{yr}\end{aligned}$$

That is, it would take almost 200,000 years to count the number of people on earth after some 800 years of population growth at a rate of 2%!

We present these numbers in part because they are patently absurd. They show how simplistic calculations with exponentials can lead to results that are mathematically correct but practically useless. We also note the effect of scale in the two figures presented, particularly in the ordinates. In Figure 8.1 we have an ordinate value of 100 billion people in 1.50 in. of graph, while in Figure 8.2 we have a value of 2,000,000 billion in 1.00 in. If we used the ordinate graph scale of Figure 8.1 to express a population of 5,000,000 billion people (the population in the year 2675 according to Figure 8.2), we would require a piece of graph paper that is 75,000 in. (or 6250 ft) long to get to that value. We can also use scaling judiciously to make exponential behavior look much more reasonable. This is illustrated in Figure 8.3, where we see a more gradual growth pattern with much smaller values for the world population. Although this curve, too, is exponential, it does not show the dramatic increases in population evident in the preceding figures.

We should also note that a change of scale is not enough to generate or dissipate exponential growth; scale changes only provide perspective by allowing more detailed examination of the exponential behavior over different ranges of values of the independent variable. What is more important is that exponential behavior can be modified and it can be used to express other kinds of response, some of which are illustrated in Figure 8.4. We see that we can obtain *decay* as well as growth, and we can obtain *asymptotic behavior*, where the response goes to a finite limiting value as the independent variable becomes infinitely large. In these cases we find that the models may be more complicated than the simple relation between growth rate and instantaneous size that was discussed earlier. While in the decay case the relation is changed only by introduction of a minus sign, in the asymptotic case we find that the exponential effect becomes less important in the neighborhood of the limiting value. All this means that exponentials show up in more ways than simple representations of unlimited growth. After presenting some mathematical results, we shall illustrate the possible ranges of behavior with a variety of examples.

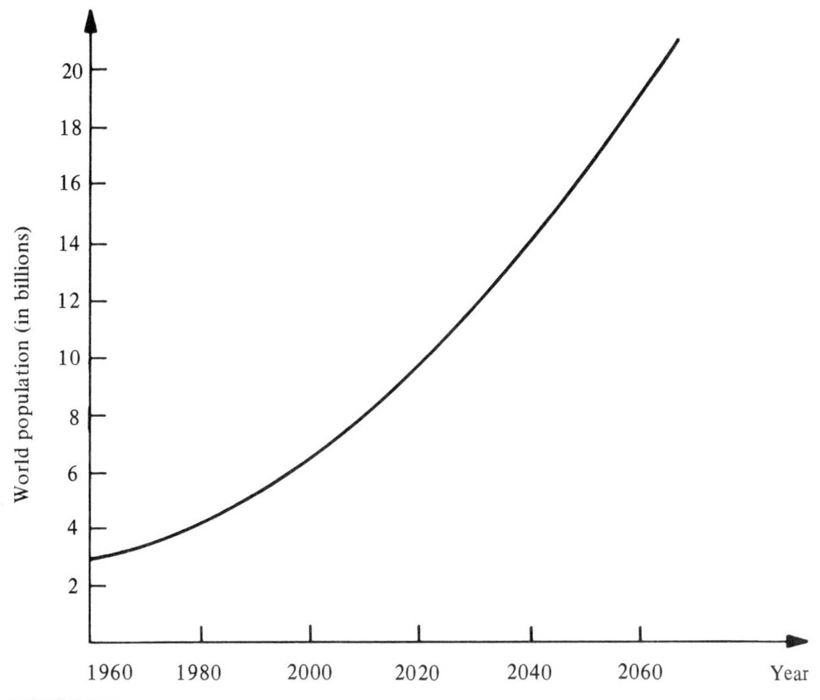

FIGURE 8.3
Short-term population projection

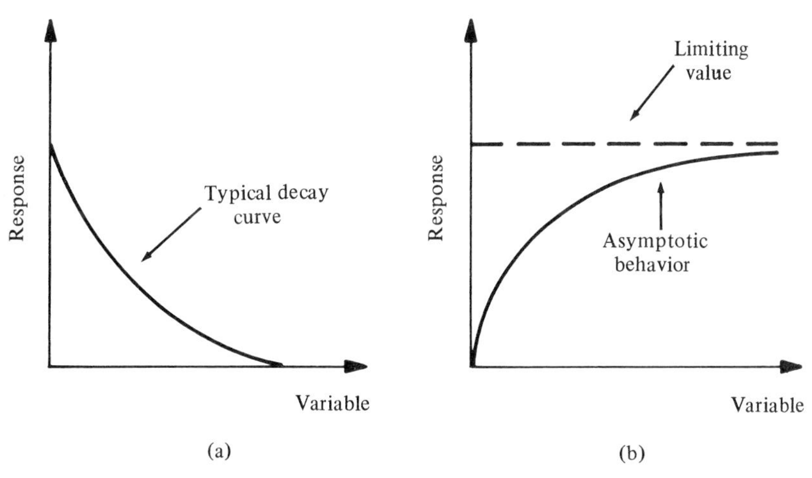

FIGURE 8.4
Alternate types of exponential behavior

EXPONENTIAL FUNCTIONS

The exponential behavior that we have discussed can be put into mathematical form as follows. Let N be the size or population of some collection of objects, let t be the independent variable (usually time) during whose variation the changes in N are measured, and let λ be a constant of proportionality. Then if the rate of change of N with t is directly proportional to the instantaneous value of N, we must have

$$\frac{dN(t)}{dt} = \lambda N(t) \tag{8.1}$$

We can see from Eq. (8.1) that the constant of proportionality λ is

$$\lambda = \frac{dN/N}{dt} \tag{8.2}$$

Thus, λ represents the fractional change dN/N of the population per unit change of time* dt. You can then verify with the usual limit process of differentiation that the solution to the first-order differential equation (8.1) is†

$$N = N_0 e^{\lambda t} \tag{8.3}$$

where the constant N_0 remains to be determined. That constant must have units of N and, since $e^0 = 1$, N_0 must be the initial population. The number e is the *base of natural logarithms* and it has the numerical value

$$e = 2.71828 \tag{8.4}$$

The solution $N(t)$ in Eq. (8.3) grows (for example, Figure 8.1) if $\lambda > 0$, decays (e.g., the curve on the left in Figure 8.4) if $\lambda < 0$, and provides an asymptotic solution like that shown in the graph on the right in Figure 8.4 when the solution is of the form

$$N(t) = N_0(1 - e^{\lambda t}), \qquad \lambda < 0 \tag{8.5}$$

Because e is the base of natural logarithms we can take the logarithm

*There are many problems in which exponential decay occurs over distance or over some other variable; however, we shall be dealing only with temporal variations in this chapter. In any event, there is no loss of generality incurred by identifying t with time.

†Also, see Appendix II.

of both sides of Eq. (8.3) to find that

$$\ln(N/N_0) = \lambda t \tag{8.6}$$

This result means that if t_c is the time required for $N(t)$ to reach a value $N(t_c) = cN_0$, then t_c can be calculated from the formula

$$t_c = \frac{\ln c}{\lambda} \tag{8.7}$$

The most frequent use of Eq. (8.7) is in the calculation of doubling times; that is, the time required for $N(t)$ to double in size. Then $c = 2$ and

$$t_2 = \frac{\ln 2}{\lambda} = \frac{0.693}{\lambda} \tag{8.8}$$

If we put λ in terms of percentage P, then $P = 100\lambda$, and the doubling time is

$$t_2 = \frac{69.3}{P} \tag{8.9}$$

In Table 8.1 we have listed some doubling times for different growth rates (percentages). This table will be useful in discussing interest rates, among other things.

We can also invert the doubling time process to calculate *half-lives* of population. Here λ would represent a (negative) decay rate, and we have $c = \frac{1}{2}$. Since $\ln \frac{1}{2} = -\ln 2$, it follows that values of time for substances or populations to reach half their initial values can be taken directly from Table 8.1 if the two headings are replaced by "Decay $P < 0$ per year (%)" and "Half-Life $t_{1/2}$ (yr)."

We also note from Eq. (8.7) another important property of exponential growth: The time t_c for the population N to grow by a constant

Table 8.1
Doubling Times for Different Growth Rates

Growth P per year (%)	Doubling time t_2 (yr)
1	69.3
2	34.6
5	13.9
10	6.93
20	3.46

factor c remains unchanged throughout the growth. Thus, from time $t = 0$ to $t = t_2$, the population doubles; from $t = t_2$ to $t = 2t_2$, the population doubles again; and so on. Thus we obtain the results in Table 8.2.

A final note on the graphical display of exponential functions is in order. We have already seen that exponential growth leads to very large numbers. Therefore, it is often convenient to use the properties of the logarithm function to "compress" the numerical values onto a *semilogarithmic plot*. This is done by plotting $\ln N(t)$ against t instead of $N(t)$ against t. From Eq. (8.3) we can calculate that

$$\ln N(t) = \ln N_0 + \lambda t \tag{8.10}$$

On a piece of graph paper whose ordinate values are expressed in natural logarithms and whose abscissa values are expressed as usual, Eq. (8.10) represents a straight line of slope λ and of intercept $\ln N_0$. In Figure 8.5 we have reproduced the data from Figures 8.1 and 8.2 in a semilogarithmic format for the time interval from 1960 to 2400.

Two final points are to be noted before we end our survey of exponential mathematics. We have seen, first, that the solution (8.3) to the differential equation (8.1) has only one arbitrary constant, in contrast to the solution (5.23) of the pendulum problem, where there are two unknown constants. This difference comes about because Eq. (8.1) has only first-order derivatives, whereas the pendulum problem is governed by an equation whose highest derivative is second order [viz., Eq. (5.18)].

The other point is that so far we have considered only elementary exponential processes, whether of growth or decay. Clearly this can often lead to impractical or absurd results, as is the case in the simplistic population projections. In the rest of this chapter we shall indicate where other models (and other limitations) can be usefully brought to bear.

Table 8.2
Growth of the Exponential Function

Time (in units of t_2)	Population $N(t)$	
$t = 0$	$N = N_0$	$= 2^0 N_0$
$= t_2$	$= 2N_0$	$= 2^1 N_0$
$= 2t_2$	$= 4N_0$	$= 2^2 N_0$
$= 3t_2$	$= 8N_0$	$= 2^3 N_0$
$= 10t_2$	$= 1024N_0$	$= 2^{10} N_0$
$= nt_2$		$= 2^n N_0$

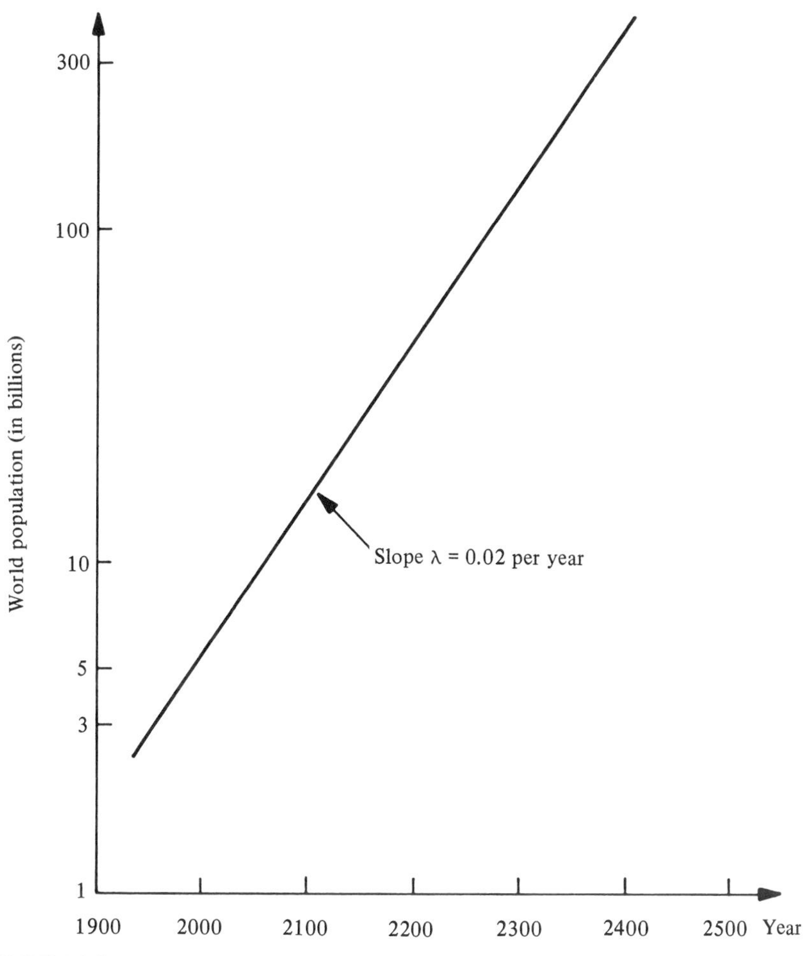

FIGURE 8.5
Semilogarithmic plot of population projections of Figures 8.1 and 8.2

RADIOACTIVE DECAY

We turn to the decay of radioactive isotopes as the first of our models on exponential behavior. It was found very early in the history of radioactivity that the activity of a radioactive material decreases with time at a rate that is an individual characteristic of each isotope. In physical processes such as alpha emission or primary beta emission, no change in the half-life can be produced by changes in pressure, temperature, chemical state, or physical environment. Thus, the half-life is a strong indicator of the identity of a radioisotope. A typical plot of the decay of a radioactive isotope would appear as in Figure 8.6. This

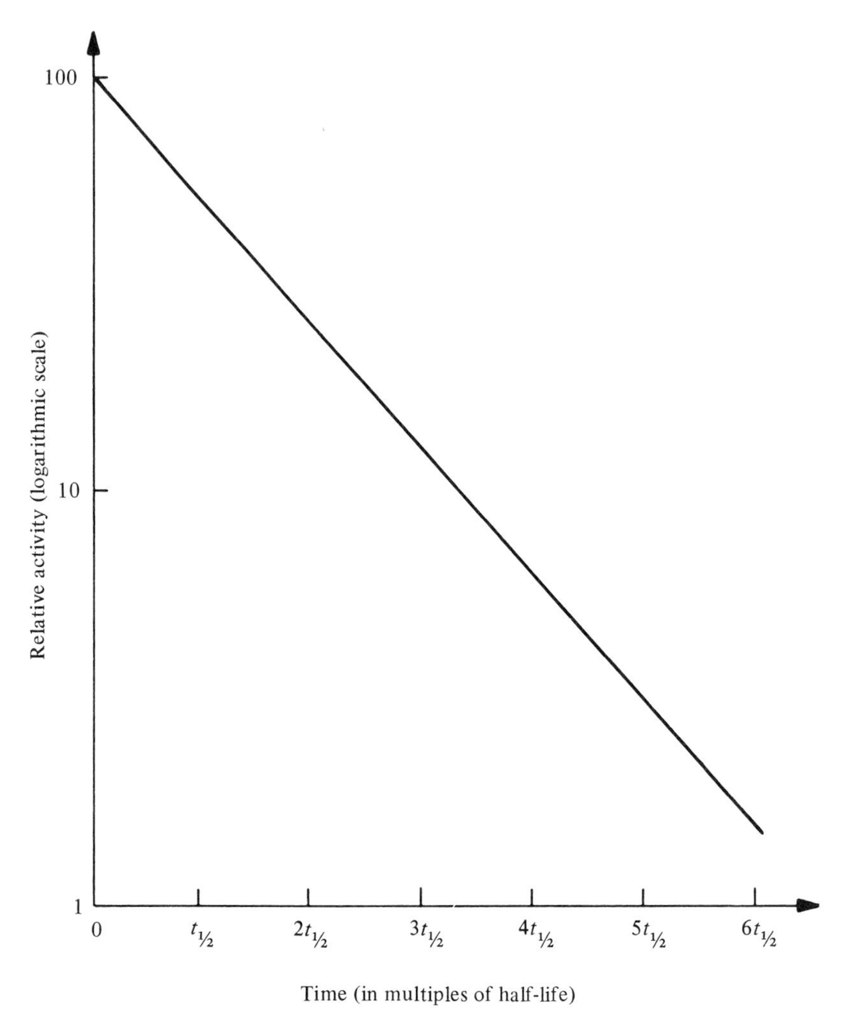

FIGURE 8.6
Semilogarithmic plot of the decay of a radioactive isotope

graph is of the same form as the plot in Figure 8.5 except that the slope of the straight line on the semilogarithmic plot is negative since λ is negative for a decay process.

The decay rates can also be used to characterize emitters as *long-lived* or *short-lived*. For example, the element thorium has a half-life of 16,500,000,000 yr, which is a long time, so it is considered a long-lived emitter. We can calculate from this half-life the fraction of atoms that decay per year, for example. From Eq. (8.7) with $c = \frac{1}{2}$ and $t_{1/2} = 1.65$

$\times\ 10^{10}$ yr, the decay constant can be calculated as

$$\begin{aligned}\lambda &= \frac{-0.693}{1.65 \times 10^{10}\ \text{yr}} = -4.20 \times 10^{-11}\ \text{yr}^{-1} \\ &= -4.20 \times 10^{-11} \frac{1}{\text{yr}} \times \frac{\text{yr}}{365\ \text{days}} \times \frac{\text{day}}{86{,}400\ \text{sec}} \\ &= -1.33 \times 10^{-18}\ \text{sec}^{-1}\end{aligned}$$

where reciprocal seconds are the units ordinarily used to express radioactive decay constants. In view of the definition (8.2) of the decay rate in terms of fractional change, this result means that only one in every $(1.33 \times 10^{-18})^{-1} = 7.51 \times 10^{17}$ thorium atoms decays in 1 sec. Or, in one year, only one of every 2.38×10^{10} atoms initially present will decay. Thus, thorium is indeed a long-lived emitter.

DISCHARGE OF A CAPACITOR IN AN RC CIRCUIT

Let us consider a simple electrical circuit wherein we apply a voltage to a loop that includes only a resistor and a capacitor. Thus, from the circuit depicted in Figure 6.12 we delete the inductor to obtain the circuit in Figure 8.7. In the absence of the inductor we can simplify Kirchhoff's voltage law as given in Eq. (6.30) to

$$V_{\mathrm{a}}(t) = V_{\mathrm{C}} + V_{\mathrm{R}} \tag{8.11}$$

In view of Eqs. (6.29a) and (6.29b) we can substitute into Eq. (8.11) to obtain the circuit differential equation in terms of the charge q:

$$R\frac{dq}{dt} + \frac{1}{C}q = V_{\mathrm{a}}(t)$$

or

$$\frac{dq}{dt} + \frac{1}{RC}q = \frac{1}{R}V_{\mathrm{a}}(t) \tag{8.12}$$

You may easily verify from Eqs. (6.29) that all the terms in Eq. (8.12) have dimensions of current (or charge per unit time or voltage over

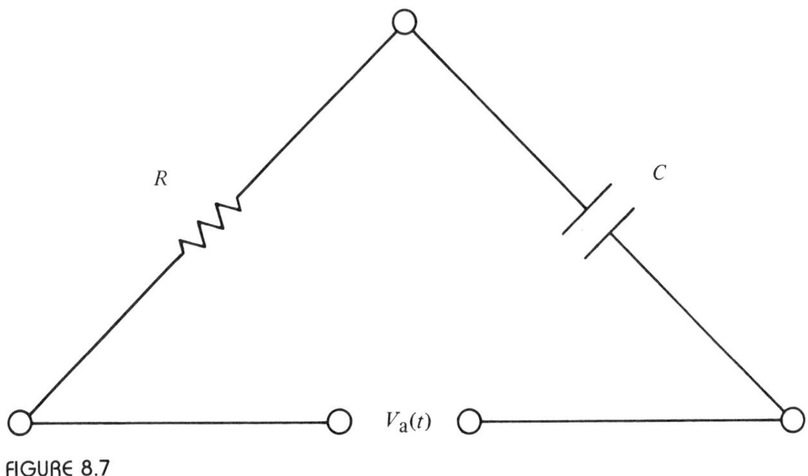

FIGURE 8.7
A simple RC circuit

resistance). Further, in the absence of an applied voltage we can write Eq. (8.12) as

$$\frac{dq}{dt} = -\frac{1}{RC}q \tag{8.13}$$

This formula is in exactly the same form as Eq. (8.1) with the decay constant given by

$$\lambda = -\frac{1}{RC} \tag{8.14}$$

The dimensions of RC are obviously dimensions of time. This may be confirmed from Eq. (8.13) or from the definitions of R and C given by Eqs. (6.29a) and (6.29b).

The solution for the behavior of the charge in time can be found to be exactly analogous to the solution (8.3), so that

$$q(t) = q_0 e^{-t/RC} \tag{8.15}$$

where q_0 is the initial charge; that is, $q_0 = q(t = 0)$. The decay is then inversely proportional to both the resistance and the capacitance, which agrees with our intuition. If the resistor is large, the discharge of current across it is slowed down; and if the capacitor is large, there will be a great deal of charge to dissipate, which can thus take considerable time.

One measure of decay rate that is often used in electrical circuit problems is the time it takes for the initial charge to be reduced to a value of q_0/e, that is, to a value of $1/e$ times the initial value. By Eq. (8.15) that time, which is called the circuit *time constant*, is clearly $t = t_c = RC$.

CHARGING A CAPACITOR IN AN RC CIRCUIT

We now consider the effect of charging a capacitor by applying a constant voltage for all time. This example is in contrast to the one just completed, where no voltage was applied, with the result that the (undetermined or previously imposed) initial charge simply decayed on its own. We take the applied voltage as a constant, $V_a(t) = V_0 =$ constant, so that Eq. (8.12) becomes

$$\frac{dq}{dt} + \frac{1}{RC} q = \frac{1}{R} V_0 \tag{8.16}$$

The solution to Eq. (8.16) is

$$q(t) = V_0 C + A e^{-t/RC} \tag{8.17}$$

which may be verified by substitution, and where A is an arbitrary constant that will be determined by assigning an initial charge to the capacitor. Since in this problem we are assigning a voltage, we choose for simplicity the case where $q(0) = 0$ so that the charge in the capacitor comes from the applied voltage V_0. With a trivial initial condition we have from Eq. (8.17) the result that

$$A = -V_0 C$$

so that

$$q(t) = V_0 C(1 - e^{-t/RC}) \tag{8.18}$$

This result is analogous to the plot on the right in Figure 8.4 and is displayed in more detail in Figure 8.8. We see that, starting from its initial zero value, the charge increases in an exponential form—although not to infinity. The charge increases to the *asymptotic value* of

$$q_\infty = q(t \to \infty) = V_0 C \tag{8.19}$$

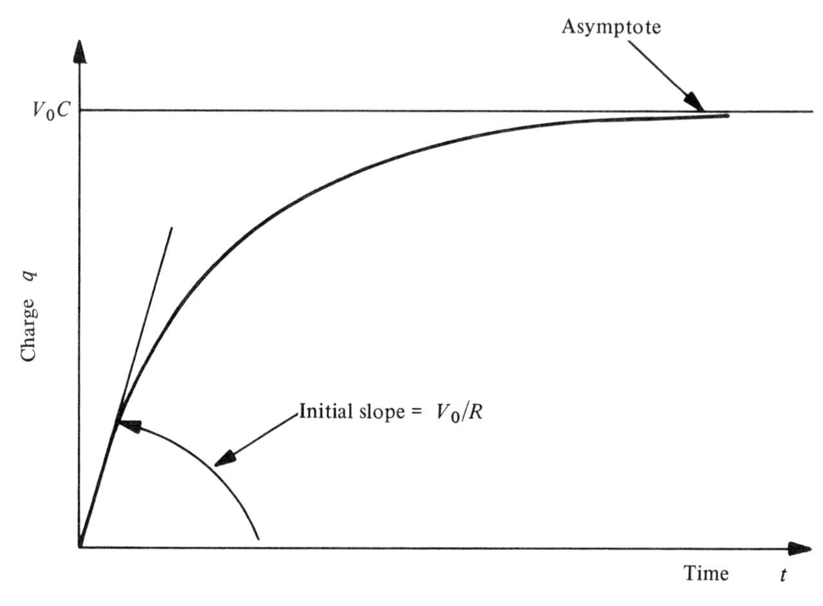

FIGURE 8.8
Charging of a capacitor in an RC circuit

which is the maximum charge the capacitor can hold for given values of capacitance C and applied voltage V_0. We also note that at $t = 0$ the slope of the charge–time curve is such that $dq(0)/dt = V_0/R$. If the slope were zero, we would have either an infinite resistance or no applied voltage. In either case, with no initial charge q_0 and no initial rate of change of charge dq/dt, we would generate no charge in the capacitor. Again, whether we have infinite resistance or no applied voltage, the result is intuitively pleasing.

A final observation here. We have charged a capacitor even though its decay constant $\lambda = -1/RC$ is negative. Thus we have imposed growth on an exponentially decaying system, which serves to point out that external conditions can influence the behavior of exponential systems as much as the sign of the constant λ. Here the external conditions were represented by a constantly applied voltage and a condition of zero initial charge.

INFLATION

Inflation is one of the major economic and political problems of our time. Its effects include the devaluation of money (and consequent price increases) as well as changes in the value of one nation's currency in comparison with another's. These economic imbalances can lead to unemployment, trade embargoes, and severe international economic dislocations. We cannot deal with problems of such magnitude here, but we do wish to make the points that inflation too is an exponential phenomenon, and that the arithmetic of inflation is provocative.

For example, gasoline that in 1978 cost \$0.70 per gallon sold for a nickel a gallon in 1933, an increase in price of \$0.65 per gallon over 45 years. The annual inflation rate can be calculated from Eq. (8.6), which here takes the form

$$\begin{aligned}\lambda &= \frac{\ln(0.70/0.05)}{45}\\ &= 0.0586 \text{ per year}\end{aligned}$$

which corresponds to an inflation rate (average) of just under 6% per year. This inflation rate has thus caused the price to go up by a factor of 14 in a 45-yr time span.

In the same way we can calculate the declining value of the dollar due to inflation. Here we would use Eq. (8.3) with a negative exponent. For example, with an annual decline of 7% in purchasing power, the value of a single dollar after t yr is

$$\text{value}(\$) = e^{-0.07t}$$

Therefore, after 1, 10, and 20 yr the dollar will be worth \$0.93, \$0.50, and \$0.25, respectively. In 66 yr of currency devaluation at a 7% annual rate, the dollar we started with would be worth only a penny!

We are not suggesting that inflation is an easy problem because it is easy to do exponential arithmetic. We do want to point out, however, that the cumulative effects of percentages in economics—in terms of inflation, salary raises, pension benefits, and the like—can be enormous. On the other hand, schemes such as *indexing* have been suggested to limit some of the consequences of inflation. Also, we have ignored in these examples the increase in purchasing power that has been a phenomenon in our society because of technical innovation and improved productivity. Suffice it to say that the politics and economics of exponential growth in monetary affairs need much further study.

COMPOUND INTEREST

After the preceding remarks on world population growth and inflation, it comes as no surprise that the compounding of interest on an investment or a loan is also exponential behavior. However, it is worth spending a few moments to see how continuous compounding of interest produces "effective" interest rates that are often considerably higher than quoted "discrete" rates. For example, a standard quotation on a credit card statement is an interest rate of 1.5% per month, which is also said to be the same as 18% per year. If the monthly interest were compounded on a monthly basis, the effective annual interest rate is found from the 12-fold multiplication

$$(1.015)(1.015) \cdots (1.015) = (1.015)^{12} = 1.1956$$

This produces an effective annual interest rate of 19.56%. If these rates are used in a continuous calculation we would find from Eq. (8.3) that

$$\begin{aligned} \frac{N}{N_0} &= e^{(0.015)(12 \text{ months})} = e^{(0.18)(1 \text{ yr})} \\ &= 1.1972 \end{aligned}$$

which shows an effective rate of 19.72% per year. That is, the effective rate due to calculating interest continuously is higher than the simple product of a monthly rate multiplied by the number of months, and even higher than the monthly compounding result. This is due to *continuous compounding.*

Another way of looking at the process of continuous compounding is to consider what happens when a person invests money at, say, 10% per year. The interest can be calculated and distributed in discrete payments of 10% annually, 5% semiannually, $2\frac{1}{2}$% quarterly, and so on. We can tabulate effective rates for each of those terms as in Table 8.3. We see from the table that the effect of compounding is to increase the effective yield. The continuous result would be

$$\frac{N}{N_0} = e^{(0.10)(1 \text{ yr})} = 1.1051709$$

which is not significantly larger than the result of compounding every day of the year. In fact, we can go through a limiting process to show that if we compounded even more often than once a day, we would have to reach the continuous effective rate just calculated. For m compoundings a year, the interest calculation is such that

$$\begin{aligned} \left(\frac{N}{N_0}\right)_m &= \left(1 + \frac{0.10}{m}\right)^m \\ &= \left(1 + \frac{1}{x}\right)^{0.10x} \end{aligned} \tag{8.20}$$

Table 8.3
Compound Interest at 10%

Number of compoundings per year	Calculation	Value of a unit investment
0	1	1.0000
1	$(1 + 0.10)^1$	1.1000
2	$\left(1 + \frac{0.10}{2}\right)^2$	1.1025
4	$\left(1 + \frac{0.10}{4}\right)^4$	1.1038
12	$\left(1 + \frac{0.10}{12}\right)^{12}$	1.1047
365	$\left(1 + \frac{0.10}{365}\right)^{365}$	1.1051559

where to arrive at Eq. (8.20) we have introduced the variable $x = m/0.10$. If the number of compoundings becomes infinitely large, then $m \to \infty$ and $x \to \infty$. Hence, in the limit,

$$\begin{aligned}\left(\frac{N}{N_0}\right)_\infty &= \lim_{x\to\infty}\left(1 + \frac{1}{x}\right)^{0.10x} \\ &= \lim_{x\to\infty}\left[\left(1 + \frac{1}{x}\right)^{x}\right]^{0.10} \\ &= e^{0.10} = 1.1051709 \end{aligned} \tag{8.21}$$

by the formal definition of the number e, that is,

$$e \equiv \lim_{x\to\infty}\left(1 + \frac{1}{x}\right)^{x} \tag{8.22}$$

GROWTH IN DEMAND FOR HIGHWAYS

We can also show that the growth in demand for highways is exponential in nature. Such a demonstration rests on three assumptions. The first assumption is that a fixed fraction r of total annual highway revenue R (from gasoline taxes, tolls, etc.) is always earmarked for highway construction. This assumption is simply a reflection of existing federal and state laws. Another assumption is that the total revenue R is directly proportional to the annual mileage* M driven by the population

*We use mileage as a generic term to indicate distance without specifying the units in which the distance is measured.

with the constant of proportionality being called m and having dimensions of revenue/mileage:

$$R = mM \tag{8.23}$$

The third assumption is the most interesting one, and it is based on empirical (and perhaps cynical) observation. That is, it seems that when a new highway is built, it is almost immediately filled to capacity. (A truly cynical restatement of this observation is that new highways are built to create or enlarge traffic jams rather than to alleviate existing traffic jams!) Then we could assume that the annual driven mileage M is proportional to the total length (in miles or some other unit of length) L of highways, the constant of proportionality being called l and being dimensionless:

$$M = lL \tag{8.24}$$

The amount (in length) of new highway construction per year dL/dt must be proportional to the portion of revenue rR expressly devoted to construction; that is,

$$\frac{dL}{dt} = \frac{rR}{C} \tag{8.25}$$

where C is the construction cost per unit length of new highway. In view of Eqs. (8.23) and (8.24) we can rewrite the right-hand side of Eq. (8.25) as

$$\frac{dL}{dt} = \frac{rmM}{C} = \frac{rmlL}{C}$$

or

$$\frac{dL}{dt} = \left(\frac{rml}{C}\right)L \tag{8.26}$$

Thus we can model the construction of new highways as an exponential process in which the rate constant is proportional to the revenue devoted to construction and the mileage driven and inversely proportional to the cost of new construction. These effects in the rate constant can be seen to be quite reasonable from our experience, so we would be inclined to accept this model as an initial approach to estimating the growth of demand for highways.

POPULATION GROWTH

In earlier parts of this chapter we discussed population growth and presented several graphs based on simple exponential growth at a uniform growth rate. For a world population growth model this means using Eqs. (8.1)–(8.3) with the rate constant λ representing the net difference between birth and death rates. For a simple exponential model of a particular state or country we might want to include the immigration and emigration rates in the calculation of λ, but the fundamentals don't change and uninhibited exponential growth would still result.

What we would like to do here is expand the notions of exponential growth to incorporate the idea of limited growth. The conceptualization of limits to growth is not new and it has received considerable attention in recent years. Such factors as renewable and nonrenewable resources, capital formation, availability of energy, food supplies, pollution, education, and family planning have been modeled by some researchers in order to demonstrate that there are in fact rather strict limits to growth. Such models are complicated and are not widely understood; they are also disputed by some. However, it is safe to say that much of the modeling involved here is exponential in that the growth rates of many of the parameters just listed do depend on their instantaneous values as well as on the values of other variables. That is, many of the growth variables are *coupled*, since the rate of growth of capital formation, for example, may depend on the indices of pollution and energy supplies as well as on the instantaneous money supply.

It is also true that many of the relations between growth rate and instantaneous values are not limited to dependence on the (linear) instantaneous value alone. That is, the right-hand side of Eq. (8.1) may be more complex. We will illustrate this with a population model into the mathematics of which we shall build a limit to the growth of population. We can go at this model construction in two ways. One way is to consider how we might replace the right-hand side of Eq. (8.1) with a function that does not itself increase indefinitely with population. If the right-hand side became zero, or even negative, then the growth rate would become zero or negative. This of course would lead to a stabilized or declining population.

For example, let us modify Eq. (8.1) to the form

$$\frac{dN(t)}{dt} = \lambda_1 N - \lambda_2 N^2 \qquad \lambda_1, \lambda_2 > 0 \tag{8.27}$$

Here λ_1 would correspond to the actual uninhibited growth rate or the potential growth rate, while the ratio λ_1/λ_2 would be the maximum obtainable population. This is true because it is the value of the population $N(t)$ for which the growth rate vanishes and the population would thus be stable at the maximum population value. That is, let $dN(t)/dt$ vanish, from which it follows that

$$\lambda_1 N - \lambda_2 N^2 = 0$$

or

$$N_{\mathrm{m}} = N_{\mathrm{max}} = \frac{\lambda_1}{\lambda_2} \tag{8.28}$$

Then we can rewrite Eq. (8.27) in terms of N_{max} and λ_1, eliminating λ_2 by virtue of Eq. (8.28):

$$\frac{dN}{dt} = \lambda_1 N\left(1 - \frac{N}{N_{\mathrm{m}}}\right) \tag{8.29}$$

This formula represents a modification of the elementary exponential model by the incorporation of a factor that is in effect the proportion of the unrealized population, the population represented by the difference between the maximum population and the instantaneous population. It can be shown that the solution to Eq. (8.29) subject to the initial condition $N(t = 0) = N_0$ is

$$N(t) = \frac{N_{\mathrm{m}}}{1 + \left(\dfrac{N_{\mathrm{m}}}{N_0} - 1\right)e^{-\lambda_1 t}} \tag{8.30}$$

For $t = 0$, Eq. (8.30) yields the appropriate initial condition, while for $t \to \infty$ we obtain from Eq. (8.30) the maximum population N_{m}. The behavior in between these limits is displayed in Figure 8.9, which is known as a *logistic growth curve*.

We indicated earlier that there are two ways (at least) to modify Eq. (8.1) to get away from simple uninhibited growth. Another way is to assume that the growth rate [on the right-hand side of Eq. (8.1)] can be expanded into a series, as

$$\frac{dN(t)}{dt} = C_0 + C_1 N + C_2 N^2 + \cdots \tag{8.31}$$

We would have to say first that $C_0 = 0$, since in the absence of any

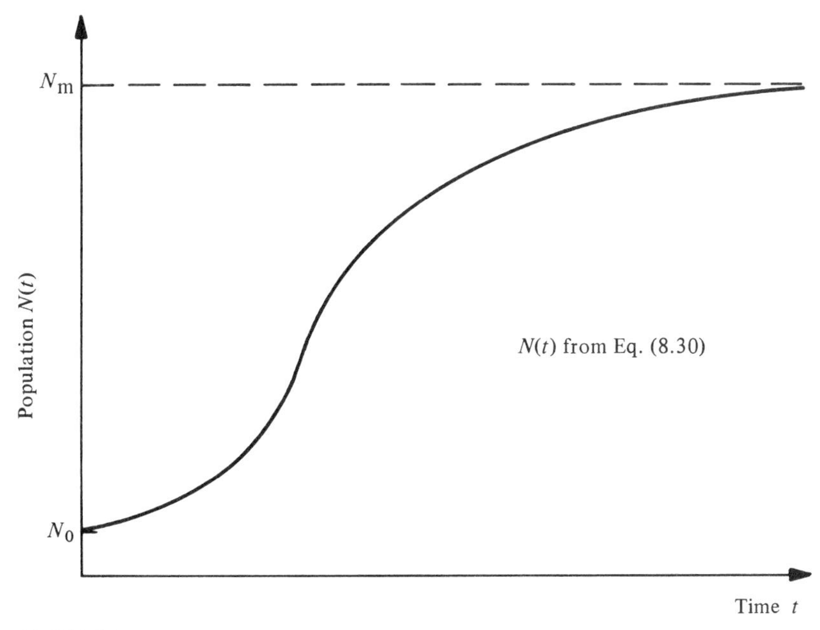

FIGURE 8.9
Population growth with finite limits

population ($N = 0$) the rate of growth must vanish. The constant C_1 would obviously be our potential growth rate λ_1. Then the constant C_2 can be evaluated in much the same way as we evaluated λ_2 in Eq. (8.27). Then we could go on and add more terms, thus adding more complexity; at the same time, we would have to impose further conditions to determine the additional constants.

We have to observe again that we have not by any means exhausted the spectrum of exponential population models. Nevertheless we have demonstrated that there are models that do not lead to unrestricted growth, and have thus opened the door to further study of exponential and population models.

THE LOTKA–VOLTERRA EQUATIONS

In this section we shall expand the logistic growth model to see what happens in a *coupled* system when two species interact in a limited way. We have seen that the inclusion of the term $-\lambda_2 N^2$ in the logistic growth model resulted in the leveling out of otherwise unrestrained growth. That is, the term $-\lambda_2 N^2$ described the *inhibition* of growth. We start here with a form of the logistic model given in Eq. (8.29). We suppose that there are two populations, host and parasite (or, equivalently, prey and predator), which we denote by H and P, respectively. We then assume that the mechanism of growth (decay) inhibition for *each* population is the reduction of its effective growth (decay) rate by a factor proportional to the size of the *other* population. This is why this system is a *coupled* system. Thus we assume the rate equations

$$\begin{aligned} \frac{dH}{dt} &= \lambda_H(1 - \alpha P)H \\ \frac{dP}{dt} &= -\lambda_P(1 - \beta H)P \end{aligned} \tag{8.32}$$

We see in Eqs. (8.32) that the parasite population decreases the growth rate of the hosts, which is what parasites do. We further see that the presence of the hosts slows the decline of the parasite population, which is also in accord with our intuition, since the absence of hosts must inevitably mean a dwindling of parasite numbers because the parasites have nothing on which to feed. Equations (8.32) are variously called the *Lotka–Volterra* equations, or the *predator–prey* or *parasite–host* equations.

The parameters λ_H and λ_P in Eq. (8.32) represent the uninhibited growth and decay rates, respectively, of the host and parasite populations. The constants α and β are determined by noting that the populations of hosts and parasites attain their *equilibrium* values (H_e and P_e, respectively) when the rates (derivatives) dH/dt and dP/dt both vanish. When this happens,

$$1 - \alpha P_e = 0 \quad \text{and} \quad 1 - \beta H_e = 0 \tag{8.33}$$

It then follows that the Lotka–Volterra equations can be written as

$$\begin{aligned} \frac{dH}{dt} &= \lambda_H\left(1 - \frac{P}{P_e}\right)H \\ \frac{dP}{dt} &= -\lambda_P\left(1 - \frac{H}{H_e}\right)P \end{aligned} \tag{8.34}$$

Again, these equations have a marked similarity to the logistic model given in Eq. (8.29). There are, however, some important differences. In addition to the obvious difference that the inhibition is due to the population of the other species, we note also that the populations at which the derivatives vanish are termed equilibrium populations rather than maximum populations. In the simple logistic model of Eq. (8.29) the equilibrium population and the maximum population are the same. Here that is not the case, since, for example, the equilibrium population of parasites may be surpassed, which in turn results in a *change in sign* of the growth rate (into a decay rate) of the host population. In other words, the parasite population may exceed its equilibrium value ($P > P_e$), so that the sign of dH/dt will change* because of the factor $1 - P/P_e$. Then we can have *oscillations* of the population sizes about their equilibrium values. We shall discuss this further in what follows.

The Lotka–Volterra equations cannot be solved in the sense that an analytical closed-form solution can be written for the functions H and P. However, we can get some information without closed-form solutions. First we note that we can divide† Eqs. (8.34), one by the other, to find that

$$\frac{dH}{dP} = -\frac{\lambda_H(1 - P/P_e)H}{\lambda_P(1 - H/H_e)P} \tag{8.35}$$

or

$$\frac{1}{\lambda_H}\left(\frac{1}{H} - \frac{1}{H_e}\right) dH + \frac{1}{\lambda_P}\left(\frac{1}{P} - \frac{1}{P_e}\right) dP = 0 \tag{8.36}$$

Equation (8.36) can be integrated, since all of its terms represent exact differentials. That is, Eq. (8.36) is equivalent to writing

$$\frac{1}{\lambda_H} d\left(\ln H - \frac{H}{H_e}\right) + \frac{1}{\lambda_P} d\left(\ln P - \frac{P}{P_e}\right) = 0 \tag{8.37}$$

If we integrate Eq. (8.37), we find that

$$\frac{1}{\lambda_H}\left(\ln H - \frac{H}{H_e}\right) + \frac{1}{\lambda_P}\left(\ln P - \frac{P}{P_e}\right) = \text{constant} \tag{8.38}$$

Equation (8.38) represents a family of closed curves on a set of axes where, say, H is the abscissa and P is the ordinate. Each member of the family corresponds to a different value of the constant on the right-hand side of Eq. (8.38). Such a family is displayed in Figure 8.10. We see that for small values of the constant, the closed curves are nearly elliptical. We can also identify the flat spots in the curves as occurring at abscissa values of $H = H_e$, when the slope dP/dH vanishes [see Eq. (8.35)]. Similarly, there are vertical tangents at ordinate values of

*Remember that by their very nature the population sizes H and P are positive numbers.

†What we are doing is eliminating the independent variable (time) and solving for the populations in terms of one another.

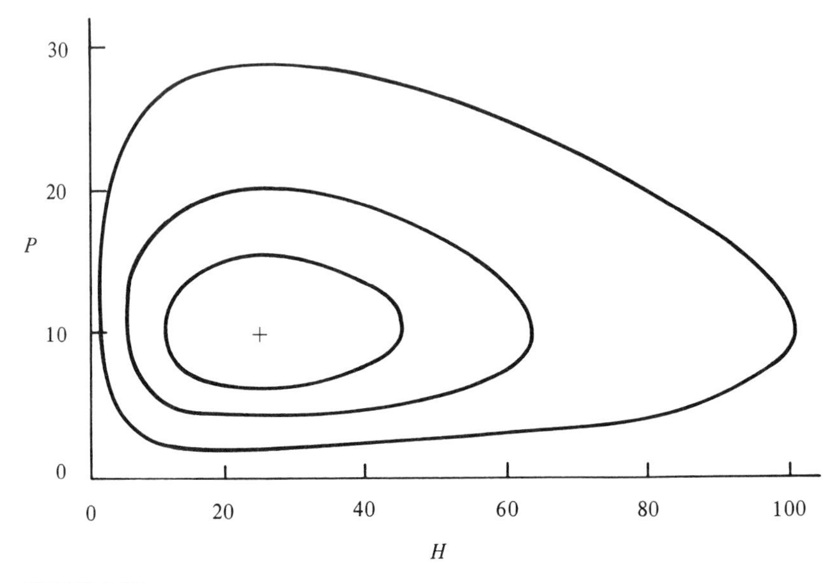

FIGURE 8.10
Three curves of the family of Eq. (8.38). Here, $\lambda_H = 1.00$ per unit time, $\lambda_P = 0.50$ per unit time, $P_e = 10$, and $H_e = 25$. (Reprinted from E. C. Pielou, *An Introduction to Mathematical Ecology*. Copyright © 1969. Used by permission of the publisher, John Wiley and Sons, Inc.)

$P = P_e$, when the slope dP/dH is infinite. As the constant increases, the resemblance to ellipses disappears. Most importantly, however, we can (for given values of the constant) trace the two populations and can thus examine their relative sizes and other characteristics as they interact.

We can also get more information by developing an *approximate solution* to the Lotka–Volterra equations. The approximation is to be found for populations that are close to their equilibrium values. That is, we shall assume that we can write the populations in terms of *small increments*—h and p, respectively—from the equilibrium populations:

$$h = H - H_e \qquad p = P - P_e \tag{8.39}$$

In terms of the small increments (or *perturbations*), the host–parasite equations are written as

$$\begin{aligned} \frac{dh}{dt} &= -\lambda_H \left(\frac{H_e}{P_e}\right)\left(1 + \frac{h}{H_e}\right) p \\ \frac{dp}{dt} &= \lambda_P \left(\frac{P_e}{H_e}\right)\left(1 + \frac{p}{P_e}\right) h \end{aligned} \tag{8.40}$$

We notice that in Eqs. (8.40) the right-hand sides contain terms that are linear in the small increment as well as products of the increments, for example, $p + ph/H_e$ in the first of Eqs. (8.40). We now *neglect* these products as being of higher order in the neighborhood of the equilibrium point. When we do this—when we neglect products of perturbation terms—we are performing a *linearization* that is similar in spirit to the linearization of the sinusoid in the pendulum model. That is, we write

$$\begin{aligned} \frac{dh}{dt} &\cong -\lambda_H\left(\frac{H_e}{P_e}\right)p \\ \frac{dp}{dt} &\cong \lambda_P\left(\frac{P_e}{H_e}\right)h \end{aligned} \tag{8.41}$$

We now differentiate the second of Eqs. (8.41) with respect to time,

$$\frac{d^2p}{dt^2} = \lambda_P\left(\frac{P_e}{H_e}\right)\frac{dh}{dt}$$

and then substitute from the first of Eqs. (8.41) for dh/dt,

$$\frac{d^2p}{dt^2} = -\lambda_P\lambda_H p$$

or

$$\frac{d^2p}{dt^2} + \lambda_P\lambda_H p = 0 \tag{8.42}$$

This is the classical oscillator equation (see Chapter 6), with a natural frequency equal to the square root of $\lambda_P\lambda_H$! Thus we have demonstrated that the Lotka-Volterra model does have within it the capacity to model oscillatory behavior about the equilibrium points. In particular, it is easy to show that a solution to Eqs. (8.41) and (8.42) is

$$\begin{aligned} p &= p_0 \cos\sqrt{\lambda_P\lambda_H}\, t \\ h &= -p_0\sqrt{\frac{\lambda_H}{\lambda_P}}\left(\frac{H_e}{P_e}\right)\sin\sqrt{\lambda_P\lambda_H}\, t \end{aligned} \tag{8.43}$$

where p_0 is a constant to be fixed by initial conditions. In terms of the original host and parasite populations we can write [see Eqs. (8.39)]

$$P = P_e\left(1 + \frac{p_0}{P_e}\cos\sqrt{\lambda_P\lambda_H}\,t\right)$$
$$H = H_e\left(1 - \frac{p_0}{P_e}\sqrt{\frac{\lambda_H}{\lambda_P}}\,\sin\sqrt{\lambda_P\lambda_H}t\right) \tag{8.44}$$

This representation of the solution makes explicit the oscillations of the two populations about their equilibrium values. It is interesting to note that the period of these oscillations, $T_0 = 2\pi/\sqrt{\lambda_P\lambda_H}$, is the same for both host and parasite!

We can also demonstrate [either directly from Eqs. (8.43) or by getting first integrals of Eqs. (8.41)] that

$$\frac{1}{\lambda_H}\left(\frac{h}{H_e}\right)^2 + \frac{1}{\lambda_P}\left(\frac{p}{P_e}\right)^2 = \text{constant} \tag{8.45}$$

Equation (8.45) is the equation for an ellipse in the coordinates h and p. Thus we have proved our earlier conjecture [see the discussion just after Eq. (8.38)] that the smaller closed curves on the H–P diagram (Figure 8.10) are very nearly elliptical.

In closing this discussion we note that we have gained considerable information about the behavior of host–parasite population systems, yet we do not have an explicit solution for the two populations. By using indirect means such as those used in developing energy integrals (or first integrals) for the pendulum models as well as the simple perturbation method, we have found that this coupled system behaves as an oscillator in the neighborhood of the equilibrium position.

LANCHESTER'S LAW

In this section we discuss another exponential model involving coupled equations for growth rates. In this model there are two populations whose decay rates are of interest, each of which is proportional to the size of the other population. Thus, because of the interaction of the two dependent variables (the two population sizes), and their respective decay rates, the variables in this problem are coupled. The model that follows was developed by a remarkable Englishman, Frederick William Lanchester,* in a successful effort to model the attrition of opposing forces at war. This model turns out, too, to be similar to the predator–

*An aeronautical engineer, Lanchester (1868–1946) was responsible for the Lanchester bomber and wrote serious works on economic problems, the theory of relativity, fiscal policies, and military strategy, as well as aerodynamics.

prey laws of mathematical biology that we developed in the preceding section, especially in that this model also has the very nice feature that we can obtain a great deal of information without solving the differential equations. In fact, this approach will also very much resemble the energy integrals of the equations of motion of the pendulum (Chapter 5).

Consider that at some time t we have populations $F(t)$ of friendly troops and $E(t)$ of enemy troops. Further, let us assume that the rates of change of these troop populations can be expressed as

$$\frac{dF(t)}{dt} = -l_E E(t)$$
$$\frac{dE(t)}{dt} = -l_F F(t) \tag{8.46}$$

The parameters l_E and l_F represent the effectiveness of the troops, enemy and friendly, respectively. The dimensions of each must be such that, for example, l_E is the loss rate of friendly troops per enemy troop per unit time. Thus, if l_E is larger than l_F, it means that the enemy troops are more effective than the friendly troops.

We now see more specifically, in Eqs. (8.46), what is meant by the term coupled, in that dF/dt depends on $E(t)$, and dE/dt depends on $F(t)$. Hence, this pair of equations is said to be a coupled set. The formal solution to these equations can be found in terms of hyperbolic functions, which are also exponentials [see, for example, Eq. (3.10)]. That solution is

$$F(t) = F_0 \cosh \alpha t - \frac{1}{\alpha} l_E E_0 \sinh \alpha t$$
$$E(t) = E_0 \cosh \alpha t - \frac{1}{\alpha} l_F F_0 \sinh \alpha t \tag{8.47}$$

where the parameter α is given by

$$\alpha^2 = l_E l_F \tag{8.48}$$

and F_0 and E_0 are, respectively, the initial sizes of the friendly and enemy troop populations, that is, the populations at $t = 0$.

With the explicit solutions of Eqs. (8.47), together with numerical values for l_E, l_F, E_0, and F_0, we can compute $E(t)$ and $F(t)$ for all values of time. Thus, we can determine which troop population will be eliminated first [$E(t) \to 0$ or $F(t) \to 0$], and thus which set of troops will win. However, we can also get this information without using the explicit solutions. Note that if we multiply the first of Eqs. (8.46) by $l_F F$ and the second by $l_E E$, we get

$$l_F F \frac{dF}{dt} = -l_E l_F FE$$

and

$$l_E E \frac{dE}{dt} = -l_F l_E EF$$

from which it follows that

$$l_F F \frac{dF}{dt} = l_E E \frac{dE}{dt} \tag{8.49}$$

Equation (8.49) is equivalent to the statement that

$$l_F \frac{d}{dt}\left(\frac{1}{2} F^2\right) - l_E \frac{d}{dt}\left(\frac{1}{2} E^2\right) = 0$$

or

$$l_F F^2 - l_E E^2 = \text{constant} \tag{8.50}$$

We determine the value of the constant in Eq. (8.50) in the same way that we found the value of the pendulum energy in Eqs. (5.31). That is, since Eq. (8.50) is a constant for all time, it must be the same constant at $t = 0$ when $E(0) = E_0$ and $F(0) = F_0$. Therefore, it follows that

$$l_F F^2 - l_E E^2 = l_F F_0^2 - l_E E_0^2$$

or

$$l_F(F_0^2 - F^2) = l_E(E_0^2 - E^2) \tag{8.51}$$

Equation (8.51) is *Lanchester's square law*. From the square law we can compute the final value of the size of the winning army when the other is annihilated. For example, if the enemy forces are reduced to a final value of zero, $E_{\text{final}} = 0$, then the remaining troops in the friendly forces amount to the number

$$F_{\text{final}}^2 = F_0^2 - \frac{l_E}{l_F} E_0^2 \tag{8.52}$$

Thus, even in victory, the number of friendly troops is reduced by an amount proportional to the square of the initial size of the enemy force. This dependence on the square of the size of the army yields some other interesting features. To illustrate the importance of the squared terms in the Lanchester result, we will consider two different scenarios for the square law. We will observe two armies of equal effectiveness, where $l_E = l_F$, for which we have the simpler result that

$$F_0^2 - E_0^2 = F^2 - E^2 = \text{constant} \tag{8.53}$$

We also postulate for these scenarios an alternate, imaginary "linear law" where

$$F_0 - E_0 = F - E = \text{constant} \tag{8.54}$$

In the first scenario, a friendly army of 50,000 soldiers meets sequentially armies of 40,000 and 30,000. In the second scenario, the friendly army of 50,000 meets the combined army of 40,000 + 30,000 = 70,000. For the linear law of Eq. (8.54), we see quickly that both scenarios result in the same outcome, that is, the friendly army will be overwhelmed by the difference of 20,000 troops.

For the Lanchester model, we see that there is a tremendous difference in the outcomes in the two scenarios. In the sequential scenario, the first engagement of 50,000 against 40,000 will result in a friendly victory with a margin of 30,000 $[(50{,}000)^2 - (40{,}000)^2 = (30{,}000)^2]$, which is enough to force a draw with the second army. If the two armies combine, however, the friendly army will easily be defeated $[(50{,}000)^2 - (70{,}000)^2 \cong -(48{,}990)^2]$. This example serves not only to point out the nature of the square law; it also indicates the importance of tactics. For, as we saw in the first scenario, the outcome against a divided army is bound to be more felicitous than against the combined force.

As a final note on the Lanchester model, we point out that in the case of a fight until one side is annihilated, we can compute the time the battle takes from Eqs. (8.47). That is, if the enemy forces are reduced to zero, we have from the second of Eqs. (8.47) that the time t_b to complete the battle is such that

$$E_0 \cosh \alpha t_b - \frac{1}{\alpha} l_F F_0 \sinh \alpha t_b = 0$$

or

$$t_b = \frac{1}{\alpha} \tanh^{-1}\left(\sqrt{\frac{l_E}{l_F}} \frac{E_0}{F_0}\right) \tag{8.55}$$

Thus, we can use the square law [Eq. (8.51)] to determine the size of the victorious army, and Eq. (8.55) to calculate the time for the battle to be completed.

Of course, the foregoing results are all predicated on the rate equations (8.46), which assumptions ought to be remembered if the model is used. This model, suitably modified to account for other effects such as the addition of reinforcements, has been used successfully to quantify the outcomes of such famous battles as that over Iwo Jima.

SUMMARY

In this chapter we have dealt with a wide variety of problems—radioactive decay, charging and discharging of capacitors, inflation, interest rates, highway demand, and population—all having the common thread of exponential behavior. Some of the exponential behavior has been exponential decay, but generally speaking, it has been exponential growth that has occupied most of our attention. We have seen how important scaling can be in evaluating and using exponential data, and we have also seen the effects of unlimited growth. In addition we have shown exponential growth or decay being modified either by superposition of an external condition, as in charging a capacitor with a superposed voltage, or by changing the nature of the growth rate equation, as in our final population model. Finally we have examined a coupled exponential system that yielded the Lanchester square law for the attrition of two armies in battle.

It is worth noting that we have touched on some very complex and very timely issues in this chapter. In no way have we "solved" any of these "problems." However, we have demonstrated that the models chosen to represent inflation growth or highway demand or whatever have a decided influence on our projections—and therefore on our perceptions of the problems and the ways we might wish to approach them in the "real world."

Problems

8.1 Show that if it takes time t_c to count the population $P(t)$, the population will increase by an amount equal to $\lambda t_c P(t)$. Assume that the population growth rate is λ.

8.2 (a) If the population counting rate is c, how long does it take to count the population at time t?

(b) How much time does it take to count the increase in population that occurred while the population at time t was being counted?

8.3 The ordinate scales of Figures 8.1 and 8.2 are, respectively, 1.5 in. = 100 billion people and $\frac{1}{2}$ in. = 1,000,000 billion people. For each of these scales, determine how much paper is required to plot the 1960 world population of 3 billion people, and the projected world population of 5.63×10^{15} people for the year 2692.

8.4 On a semilogarithmic plot, with a population of $N_0 = 3$ billion in 1960, show the population curves for growth rates of 1, 2, and 3% per year, through the year 2700. What shapes are these curves? What are their slopes and intercepts?

8.5 Verify that Eq. (8.5), $N(t) = N_0(1 - e^{-|\lambda|t})$, is a solution to a growth rate equation.

8.6 Find the time constant (in the manner of the time constant of an RC circuit) for a population decaying at a rate per unit time λ.

8.7 In order to leave accumulated by the year 2000 a balance of \$100,000 in a savings account that yielded $5\frac{1}{2}$% per annum, how much should a person invest in 1980? How much should have been set aside in 1960?

8.8 If there had been a steady inflation rate of (only, would you believe?) 3% per annum, what should have been the amounts invested in Problem 8.7 to have a purchasing power in the year 2000 of \$100,000 as measured in 1960 dollars?

8.9 Verify Eq. (8.45) directly by substitution from Eqs. (8.43). Find the constant.

8.10 Verify Eq. (8.45) by multiplying the first of Eqs. (8.41) by h, the second by p, and then by constructing an appropriate first integral.

8.11 Verify Lanchester's square law [Eq. (8.51)] by using the explicit solutions of Eqs. (8.47).

8.12 Two opposing forces have initial strengths of $F_0 = 10{,}000$ and $E_0 = 5000$ troops, with $l_E = l_F = 0.1$ troop per day. How long will a battle between them last? Who will win? When the loser is wiped out, how many troops will the winner have left? Plot a graph of the troop strengths during the battle.

8.13 Two opposing forces have initial strengths of $F_0 = 10{,}000$ and $E_0 = 5000$ troops. If $l_F = 0.1$, find the winner and how many troops he has left for $l_E = 0.2$, 0.5, and 1.0. What value of l_E would produce a draw?

8.14 During the battle for Iwo Jima, the initial troop sizes were $F_0 = 54{,}000$ and $E_0 = 21{,}500$, with $l_F = 0.0106$ and $l_E = 0.0544$. If neither side had added reinforcements, how long would the battle have lasted? How many troops would the victor have had remaining?

8.15 How many troops would the United States have had to start the battle with in order to end it in 28 days? Compare the U.S. losses in this and the preceding exercise.

9

OPERATIONS RESEARCH: LINEAR PROGRAMMING

Hard upon the heels of our discussion of Lanchester's law, we continue and expand our discussion of operations research. The phrase *operations research* is used to describe the application of the scientific method (observe, model, predict) to the operation and management of organizations involving people and other kinds of resources (for example, money and machinery). The activity of operations research really got under way in England, and then in the United States, during the Second World War, when the motive was the *optimization* of the allocation of scarce military resources. The application of science to warfare dates back to ancient times and involves such luminous names as Archimedes and Leonardo da Vinci. These applications of science were devoted to developing ingenious new tools of warfare, new armaments. The modern development of operations research was organized on the principle of improving the utilization of existing stocks of resources (armament) by careful application of the scientific method. Lanchester's law represents a very early effort in the direction of present-day operations research, being the first modern attempt to model the interaction of opposing armies at war.

The scope of operations research has expanded in the last 30 years, and it has been found applicable in civil as well as military settings. Some of the civil applications are to hospital management, criminal justice system operation, and a wide variety of commercial enterprises. A hallmark of operations research is that it is directed toward achieving *optimal* solutions to problems, for example, toward finding the mix of products for a manufacturer that will *maximize* profits, toward helping the manufacturer choose the right distribution of products among various outlet locations so as to *minimize* transportation costs, toward determining the *optimal* number of toll attendants so that traffic lines will not be too long in busy periods and so that there will not be too much idle (attendant) labor during slack periods, and so on. The obvious common characteristic in these examples is the emphasis on a *best* solution. Moreover, as we noted earlier, the approach is based on application of the scientific method to "real" problems, and it is heavily oriented toward the use of data gathered from field operations rather than laboratory studies.

A few cautionary notes are in order before we begin our presentation. First we point out that the optimal or best solution may depend on the perspective of the researcher or the clear definition of what the desired objective is. For example, consider the layout of three cities shown in Figure 9.1. New connecting highways are to be built between the three cities, and the question is: What is the best layout of these highways? It is clear that the configuration shown in Figure 9.1a will produce the shortest travel times between any pair of cities, whereas the configuration in Figure 9.1b will require the smaller amount of road construction, so it will be cheaper to build. We see that we must decide on the appropriate objective; that is, is it the cheapest system that is desired (the perspective of the taxpayer), or the one that minimizes the travel time between cities for both people and goods (the perspective of the individual traveler).

Still, the choice may not be that simple. The reduction of travel time between any two cities may in fact lead to tangible economic benefits, the benefit deriving from the more rapid delivery of merchandise or the easier access of clients, and so on. If a suitable calculation of these benefits can be made, it can be used to decrease the net costs of the first configuration, thus making the choice less clear-cut. Clearly, in order to do such calculations we must think about assigning economic values to waiting time, to increased access, and to other aspects of behavior that are not always quantifiable in an obvious way. Yet this is a process in which we engage almost every day. When we decide to drive a scenic route rather than a more direct route, we are making an economic choice (as well as a value judgment) that the cost of increased time on the road is balanced by the personal benefit of being able to view the scenery. There are really two ideas involved here, the obvious one being the *trade-off* of *costs* against *benefits*. The other idea is that, consciously or unconsciously, we are making explicit our values and preferences in such a *cost–benefit analysis*. This, too, is the stuff of operations research.

Another point that we ought to keep in mind is that, unlike much of our other modeling work, we are here modeling a decision-making process rather than a physical or other (for example, population) system. We are interested here in finding optimal strategies, in choosing among alternatives, in allocating resources. The questions we face are

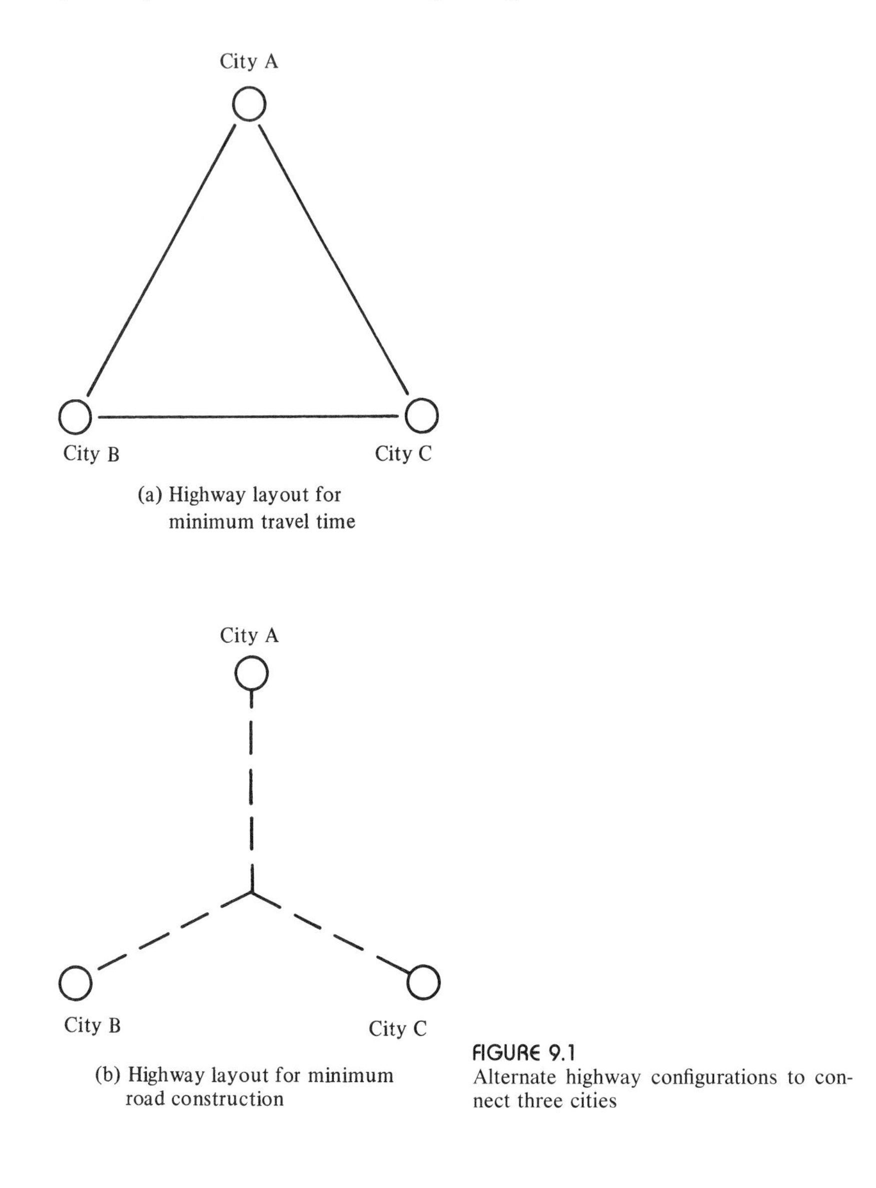

FIGURE 9.1
Alternate highway configurations to connect three cities

"How?" rather than "Why?" Thus it is not physical modeling we are doing, but mathematical modeling of some of the elements of decision making. Still, some of the same fundamentals apply. For example, as we see in the conversion of scenic benefit to economic value, we must use a consistent set of dimensions. In the same way, scaling will be important, since we may find that certain costs (or benefits) are insignificant in some scalings. Thus the modeling will be different, but much of the approach ought to be familiar.

In what follows we shall illustrate the process of operations research through the fundamental *linear programming* problem. First we shall discuss optimization in its continuous form, using the notions of extrema from the calculus. Then we shall introduce the idea of linear programming—perhaps the most fundamental methodology of operations research—in a very simple setting, and we shall apply it to a *product-mix problem*. We shall then go on to consider the costs in the *transportation problem*, and we shall show that linear programming can be applied in this context. Then we shall introduce *network analysis* and show that it too can be cast into linear programming form.

OPTIMIZATION VIA THE CALCULUS

In this section we shall outline a simple optimization problem in which we shall use classical methods of calculus, and where we shall progressively build a series of *constraints* that will limit the values that the independent variables can take. Consider first the simple function

$$U(x) = \tfrac{1}{2}x^2 - x \tag{9.1}$$

which we have plotted in Figure 9.2. We wish to find the value(s) of x for which $U(x)$ is a minimum. We see from the form of $U(x)$ that it is a parabolic function of x, so there is only one minimum, a *global* minimum. We can find the value of x for which $U(x)$ is a minimum by setting the first derivative of $U(x)$, the slope, to zero, so

$$\frac{dU(x)}{dx} = x - 1 = 0 \tag{9.2}$$

from which it easily follows that

$$x_{\min} = 1, \qquad U_{\min} = U(x_{\min}) = -\frac{1}{2} \tag{9.3}$$

Further, the derivative (or rate of change) of the slope of $U(x)$ is

$$\frac{d^2U(x)}{dx^2} = 1 \tag{9.4}$$

which is always positive. Thus the slope is always increasing as x goes from $-\infty$ to $+\infty$, so it can go through only one flat spot. There can be only one minimum, therefore, and it must be a global minimum.

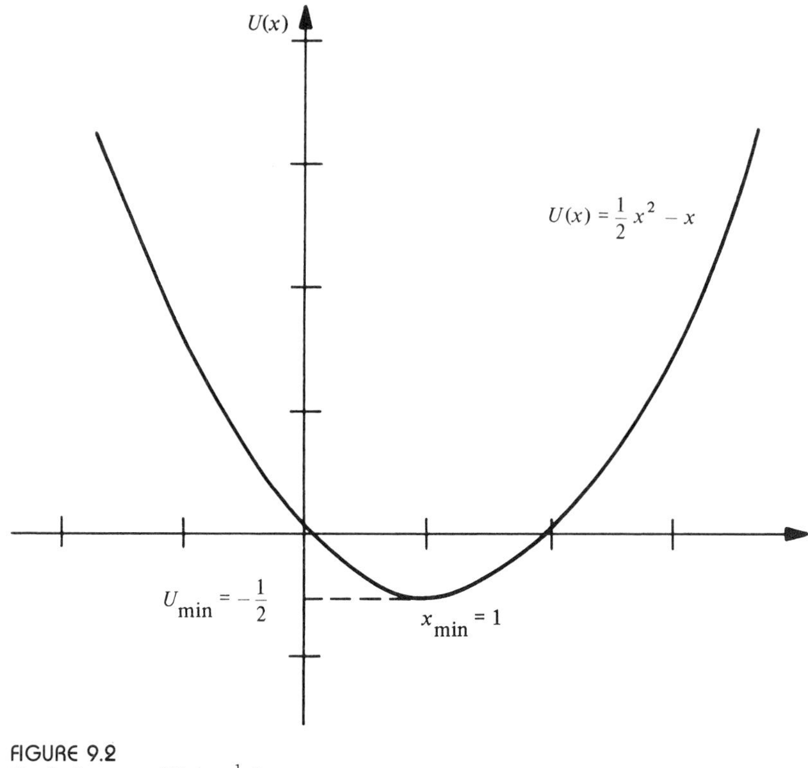

FIGURE 9.2
Parabolic curve $U(x) = \frac{1}{2}x^2 - x$

What happens if the values of the independent variable are constrained? Suppose we insist, for example, that $x = x_0$. Does it make sense to search for a minimum of $U(x)$ when the independent variable is constrained to take on the fixed value $x = x_0$? In this simple problem the answer is clear: If $x = x_0$, then perforce $U = U(x_0)$, and that point is the *constrained minimum* of $U(x)$. This constrained value is certain to be greater than the unconstrained minimum $\left(U_{min} = -\frac{1}{2}\right)$ unless $x_0 = 1$, since the unconstrained problem has only one (a global) minimum. Note, too, that this constrained minimum occurs at the intersection of the curve $U(x)$ with the line representing the constraint "boundary," $x = x_0$.

Finally, let us consider the imposition of a constraint $x \leq x_0$, that is, the independent variable x is restricted to be less than or equal to the constant x_0. Here we must do a search to find the minimum of $U(x)$ for all the *admissible values* of x: $x \leq x_0$. We can visualize this procedure by putting a line $x = x_0$ on the same graph as the curve $U(x)$, and then we can "move" this line as shown in Figure 9.3. The constraint is

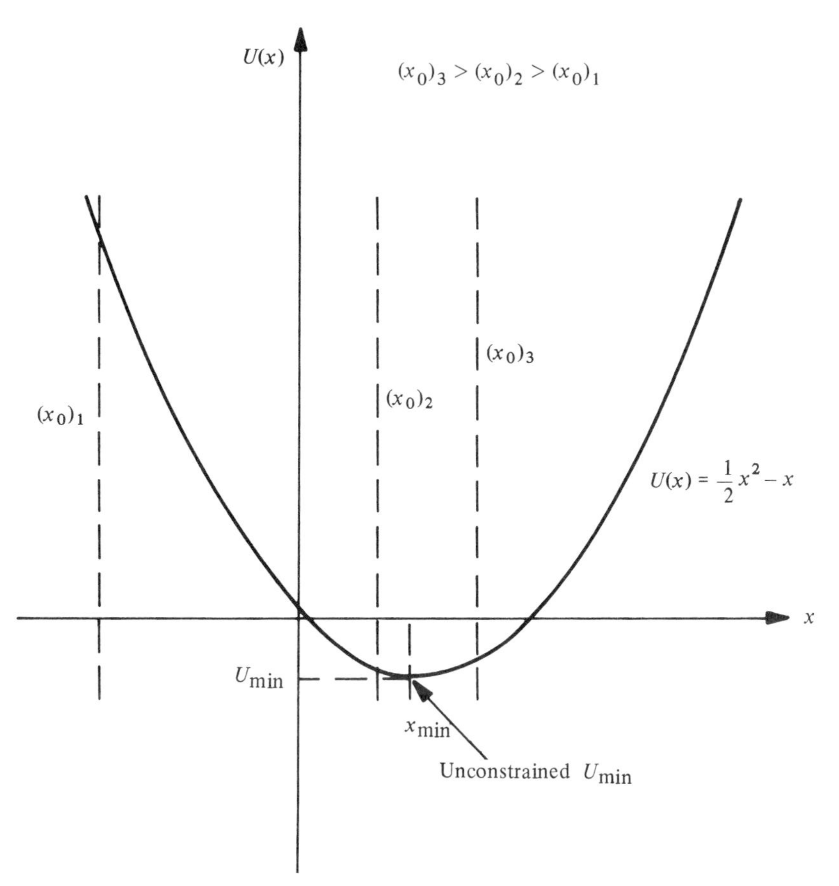

FIGURE 9.3
Parabolic $U(x)$ with constraints $x < (x_0)_i$

shown as a series of lines, $(x_0)_1 < (x_0)_2 < (x_0)_3$, so we can consider briefly the three problems of finding the minimum of $U(x)$ with $x \leq (x_0)_i$, $i = 1, 2, 3$. In the first case we have so restricted the admissible range of x that we will find a constrained minimum that is apparently much greater than the unconstrained minimum $U_{\min} = -\frac{1}{2}$. For example, if $(x_0)_1 = -3$, clearly the constrained minimum will be $U((x_0)_1) = U(-3) = 7\frac{1}{2}$. Then, as the constraint boundary moves to the right, we move toward the unconstrained minimum point $\left[\text{for } (x_0)_2 = \frac{1}{2}, U((x_o)_2) = -\frac{3}{8}\right]$ until we finally go through it. The region of *feasible solutions* for the minimum of $U(x)$ may thus include the unconstrained minimum $U_{\min}$, or it may not, depending on where the constraint boundaries happen to be.

We note that in the first of this series of elementary problems all values of x were admissible, and so all values of $U(x)$ were feasible solutions to the problem of finding the minimum value of $U(x)$. In the second case, where the *equality constraint* $x = x_0$ was imposed, the range of feasible solutions was restricted to those found at the intersection of the constraint boundary $x = x_0$ and the curve $U(x)$. Finally we imposed an *inequality constraint* $x \leq x_0$, and here the range of feasible solutions was bounded at the right end by the equality $x = x_0$, but it included the interior region $x < x_0$.

We can also see from Figure 9.4, where we have plotted some expanded schematics of the foregoing processes, that the numerical scaling of variables can be very important in the actual searches for minima, both unconstrained and constrained. If the *objective function* (that is, the function we wish to optimize) $U(x)$ varies very rapidly, as depicted in Figure 9.4a, we can see that unless the constraint boundary $x = x_0$ is very close to the location of the unconstrained minimum, we will find that the constrained minimum is significantly greater than the unconstrained minimum U_{min}. On the other hand, if the objective function $U(x)$ is relatively flat, as in Figure 9.4b, the influence of the constraint may be significantly smaller. Thus the length over which $U(x)$ varies significantly is an important scale in optimization problems.

The preceding collection of problems is an elementary one. If, for example, we wished to find the minimum of a function of two variables, x and y, say

$$U(x, y) = x^2 + 2(x - y)^2 + 3y^2 - 11y \tag{9.5}$$

we would note first that to plot $U(x, y)$ against x and y we need a three-dimensional "graph," for $U(x, y)$ is a surface rather than a curve. We would find that an unconstrained minimum occurred at the point $(x = 1, y = 1.5)$, for which $U_{\text{min}} = U(1, 1.5) = -8.25$. Even more interesting is what happens when a constraint is imposed (Figure 9.5). If we wish to find*

$$\begin{aligned} &\text{minimum} \quad U(x, y) = x^2 + 2(x - y)^2 + 3y^2 - 11y \\ &\text{subject to} \quad x + y = 3 \end{aligned} \tag{9.6}$$

we would use the calculus of several variables to find that a minimum occurs at the point $(x = 31/24, y = 41/24)$, and that $U(31/24, 41/24) = -8.02$. Note that this point is on the boundary plane $(31/24 + 41/24 = 72/24 = 3)$ that is represented by the constraint equation, and it is of

*In a three-dimensional plot of $U(x, y)$ against x and y, $U(x, y)$ represents a parabolic surface, and $x + y = 3$ represents a plane parallel to the U axis. See Figure 9.5.

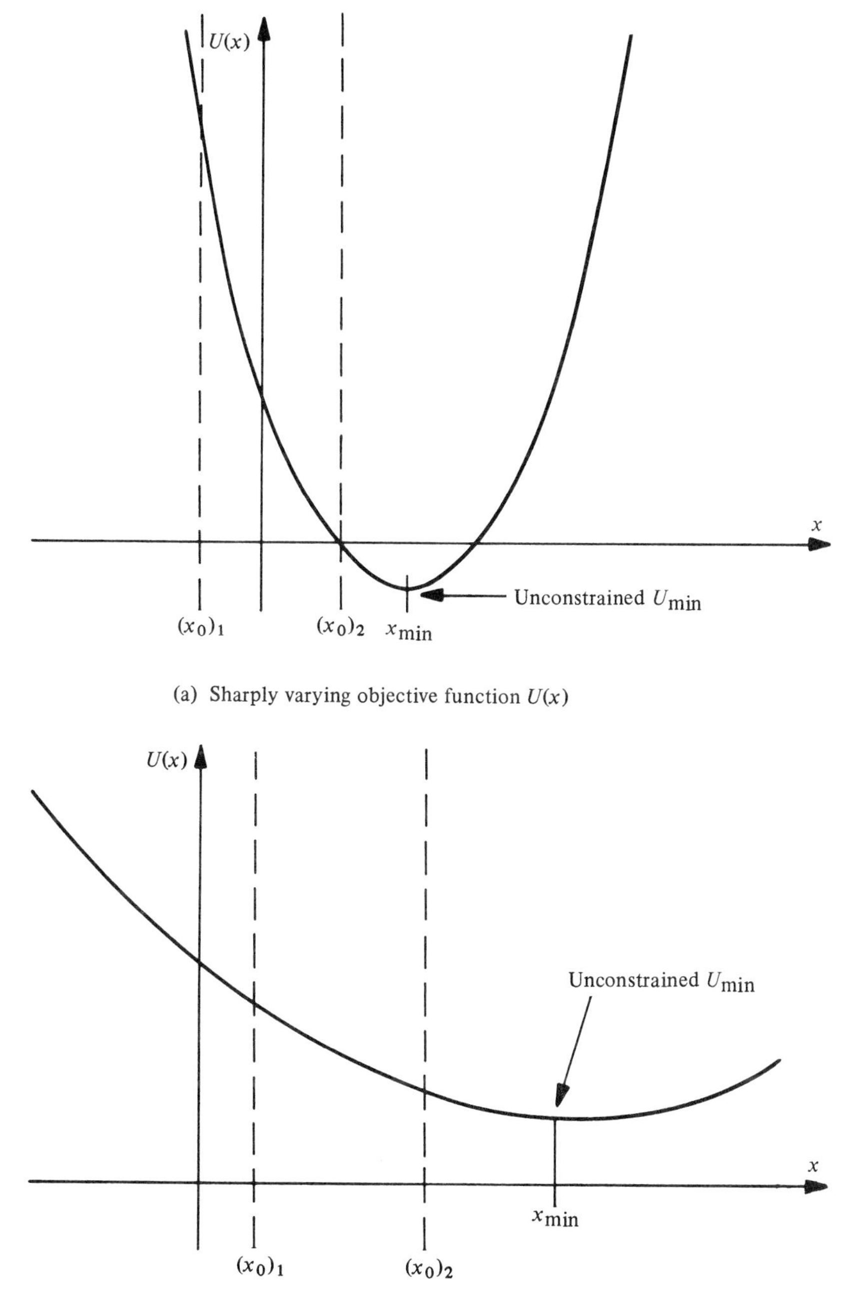

FIGURE 9.4
Schematic objective functions and boundary constraints

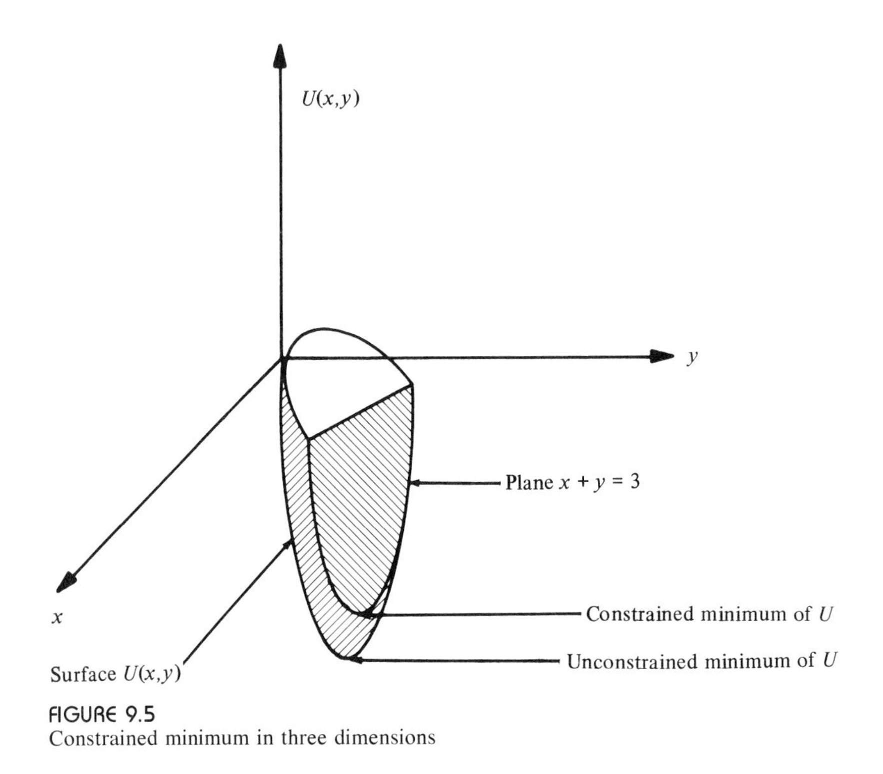

FIGURE 9.5
Constrained minimum in three dimensions

course also on the surface $U(x, y)$. So, again, we have found an optimum point on a boundary of the original problem, that boundary being the plane representing the constraint condition.

Further, we could change the equality constraint of Eq. (9.6) to the inequality constraint $x + y \leq 3$. In this case we would search on the surface $U(x, y)$ but inside the values represented by the plane $x + y = 3$. Again, we search inside and on the boundary plane representing the constraint condition, and we find here that the extremum sought is inside the intersecting boundary plane.

OPTIMIZATION VIA LINEAR PROGRAMMING

In the preceding section we saw that a search for an optimum value of a function in the presence of an inequality constraint requires a search over the interior of the region defined by the constraint boundary. Thus, in Figure 9.3, we must search for all values of $x \leq (x_0)_i$. This is because an objective function can fluctuate, and as the simple function

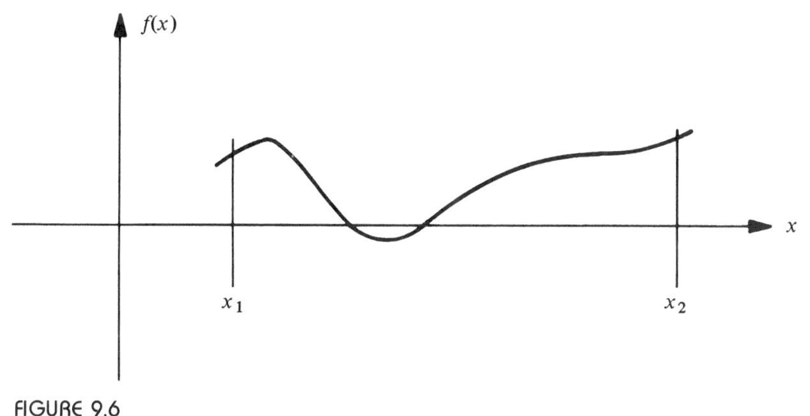

FIGURE 9.6
Schematic of admissible region of varying function

$U(x) = \frac{1}{2}x^2 - x$ displayed in Figure 9.3 does once in the interval $x \leq (x_0)_3$, it may go through several local extrema in the admissible range of x. For example, to find the maximum of the function shown in Figure 9.6 in the interval $x_1 \leq x \leq x_2$, a search of the interior region is required, and for such an elementary function the methods of the calculus are adequate. There is, however, a class of problems for which a search of the interior region is not required because the optimum point must be at one of the boundaries. This is the class of problems where the objective function is a linear function of the independent variables.

Consider for a moment the problem of finding (Figure 9.7) the

$$\begin{aligned} &\text{minimum} \quad U(x) = mx + b \\ &\text{in the interval} \quad x_1 \leq x \leq x_2 \end{aligned} \tag{9.7}$$

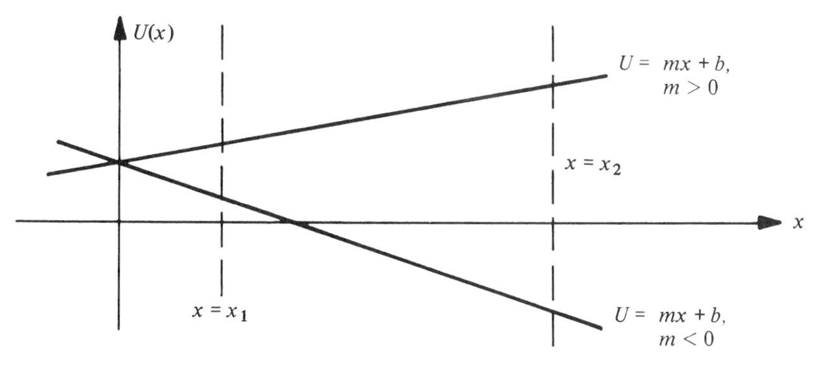

FIGURE 9.7
Schematic of a linear optimization problem

We have here two inequality constraints that define an admissible region within which the feasible solutions must lie. Notice, however, that for these linear objective functions $U(x) = mx + b$, the extrema must lie at the boundaries whether m is positive or negative. That is, for $m > 0$, U_{max} must lie on the line $x = x_2$, and U_{min} must lie on $x = x_1$. Therefore, for this linear optimization problem the interior region need not be searched, for we know *a priori* that the extrema that are sought must lie in a boundary line.

This brings us to the class of linear optimization problems that are called *linear programming* problems. These are characterized by linear objective functions and linear inequality constraints. Let us consider an example of a linear programming model. Suppose we are about to go into the business of making desks and tables, and that the desks and tables are made of oak and maple. We know how much of each kind of wood goes into each product: A desk requires 6 board-feet each of maple and of oak, while a table requires 3 board-feet of oak and 9 board-feet of maple. The lumber mill will supply up to 1200 board-feet of oak at $3.00 per board-foot, and up to 1800 board-feet of maple at $2.00 per board-foot. We also know that the market is such that we expect desks to sell at $45.00 each and tables at $42.00 each. The question that we wish to pose, and to answer, is: How many desks and how many tables should we make to assure the maximum profit?

The conditions of this problem are such that the profit earned by selling a desk is the same as that earned by selling a table [from the cost of materials for each; (see Eq. (9.10)]—that is, $15.00 each. Suppose for a moment that this is not the case, and that the profit earned by selling a table is only $9.00. Then it would seem reasonable to suppose that we can maximize the total profit by first manufacturing only desks. However, the supply of oak will be exhausted after 200 desks are made, and then there will be an excess (and presumably unusable) supply of maple that must be dealt with. Further, this production scheme also yields a total profit ($3000.) that is smaller than the maximum possible profit. That optimal profit is found through a *trade-off* of the profit obtained by making only desks in favor of the profit from a combined objective that maximizes the profit from making both desks and tables. We shall also see that by considering the alternative of making both desks and tables, as opposed to the idea of making only desks, we shall use up all the lumber. The point here is that our optimization scheme requires us to consider alternative solutions, and to choose or trade off between feasible solutions to obtain an optimal result.

We can ascertain the appropriate numbers of desks and tables by solving a linear optimization or linear programming problem. Let x be the number of desks to be made and let y be the number of tables. The objective function that we wish to maximize is the profit, the difference between income and costs. The income is due to the sale of desks and tables, and if we assume that all products made are indeed sold, we have

$$\begin{aligned}\text{income} &= (\text{price per desk}) \cdot (x) + (\text{price per table}) \cdot (y) \\ &= 45x + 42y \end{aligned} \tag{9.8}$$

The cost is reckoned in terms of the board-feet of the different woods used in each product and in terms of the cost of each type of wood; that is,

$$\begin{aligned}\text{cost} &= (\text{cost of oak})[(\text{oak/desk}) \cdot (x) + (\text{oak/table}) \cdot (y)] \\ &\quad + (\text{cost of maple})[(\text{maple/desk}) \cdot (x) + (\text{maple/table}) \cdot (y)] \\ &= 3[6x + 3y] + 2[6x + 9y] \\ &= 30x + 27y \end{aligned} \tag{9.9}$$

Thus, the objective function is

$$\text{maximize} \quad (45x + 42y) - (30x + 27y)$$

or

$$\max \quad 15x + 15y \tag{9.10}$$

Note that all the costs and profits are expressed in a common dimension, a unit of currency that happens to be the dollar in this example.

The constraints in this problem are derived from the condition that only a limited amount of wood is available from the lumber mill. The amounts of oak and maple used must be no greater than that which is available:

$$\begin{aligned}\text{amount of oak used} &= (\text{oak/desk}) \cdot (x) \\ &\quad + (\text{oak/table}) \cdot (y) = 6x + 3y \leq 1200 \end{aligned} \tag{9.11a}$$

and

$$\begin{aligned}\text{amount of maple used} &= (\text{maple/desk}) \cdot (x) \\ &\quad + (\text{maple/table}) \cdot (y) = 6x + 9y \leq 1800 \end{aligned} \tag{9.11b}$$

A final constraint enters by virtue of simple physics. The number of products made must obviously be positive, so we have a *nonnegativity constraint*,

$$x, y > 0 \tag{9.12}$$

This constraint is generally part of linear programming formulations, largely because the variables in programming problems are generally greater than zero by their very nature. It is also easy to convert any "negative" variables to "positive" by suitable changes of sign of those variables in the objective function and in the inequality constraints. Finally, restricting all variables to be nonnegative makes for easier and more efficient solution processes.

We find the solution to this linear programming problem by graphical means. In Figure 9.8 we have plotted the objective function [Eq. (9.10)] with dotted lines, and the two inequality constraints [Eqs. (9.11)] with solid lines, in the quadrant $x, y \geq 0$. The feasible region, wherein a solution is possible because all the constraints are satisfied, is shaded. We can then imagine the objective function as not one but a series of dotted lines, all of the slopes determined by the term $15x + 15y$, moving outward from the origin in the indicated direction of the arrow. As the dotted line moves through various positions in that

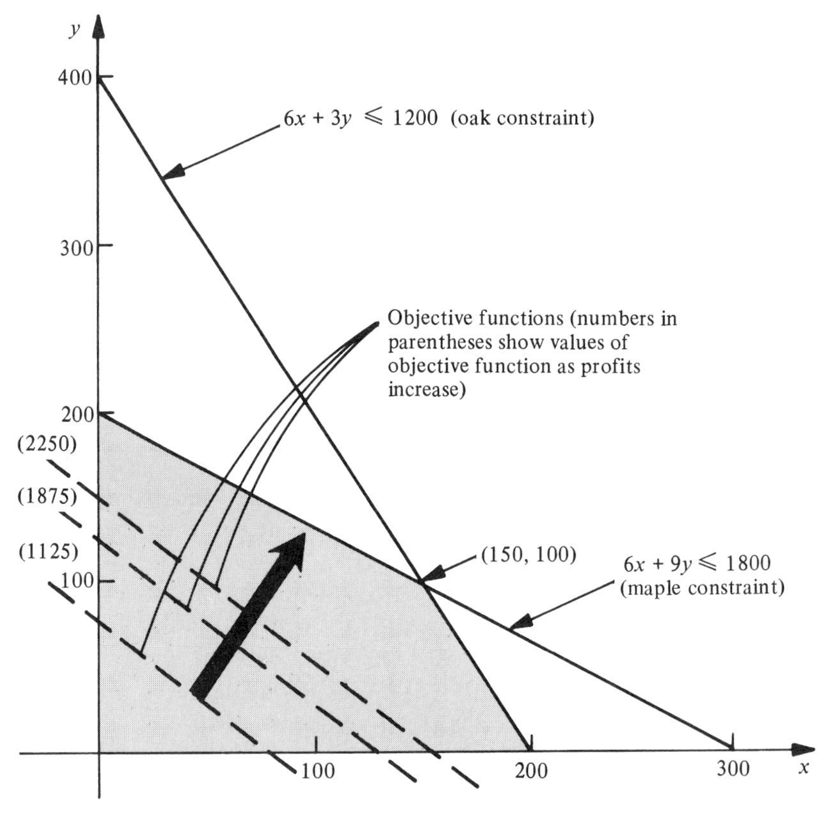

FIGURE 9.8
Graphical solution for table/desk manufacturing problem

direction, the objective function is being increased. When the dotted line moves through the point (150, 100), the objective function reaches its maximum because it is the last point in the feasible region that is also on the boundary—here, in fact, at the intersection of two (inequality constraint) boundaries.

We could have spotted this outcome immediately on the basis of the earlier discussion indicating that for linear objective functions the extremum must always lie on the boundary point. We see in the present context that for this to be true, as the line $15x + 15y$ moves from the origin, the maximum of the objective function must lie on at least one of the points (200, 0), (150, 100), or (0, 200). The first and last of these are the vertices of the boundary lines that bound the feasible region, while the middle point is the boundary intersection point. Clearly, the point among these three that maximizes the objective function $15x + 15y$ is the intersection point. So, to maximize profits we should make (and sell) 150 desks and 100 tables, for a profit of $3150, with all the lumber being used.

This concludes the presentation of a simple example of linear programming. We note that for more than two variables, and problems can be formulated with thousands of variables, the graphical approach becomes inapplicable. There exist a variety of approaches to solving the linear programming problem, notably the *simplex* method, but we shall not go into them here. Suffice to say that if an optimization problem can be formulated as a linear programming problem, the techniques for solving it exist. Also, as we shall indicate again in the next section, there are many other applications of linear programming, including the classic *feed-mix* and *product-mix* problems (which have many industrial applications) and the transportation problem.

To close this section we shall only note that there are other so-called programming problems and methodologies designed to handle more complex optimization problems. For example, the technique of *nonlinear programming* is designed to handle nonlinear objective functions. The *dynamic programming* method was developed to handle problems in which sequential, hierarchical decisions are important. In the simple analysis just completed, for example, the lumber prices might vary during the production run, so the optimum point could also change with time because different production decisions may be warranted. Finally, *integer programming* is used in problems where the variables cannot be treated as continuous, but must be treated as integers. In fact, in the example just treated there was no guarantee that our answers would be integral numbers. If they had turned out otherwise, if the number of desks had turned out to be 150.7, then we could have rounded off the result and accepted a possibly nonoptimal solution, or we could have used integer programming. The real advantages of integer programming, however, lie in applications where the variables to be manipulated are treated in a binary (for example, zero or one) fashion. In *scheduling* problems, for instance, as in ordering airline routes between cities, a plane either flies a particular route or it doesn't. Integer programming is most useful in such "go or no go" situations.

THE TRANSPORTATION PROBLEM

We wish to present here another classical problem in the field of operations research, the *transportation problem*; and although we shall not discuss solution techniques, we shall see that the transportation problem can be formulated as a linear programming problem. Further, it also falls into the category of *network problems*, a topic to which we shall return. The transportation problem we shall look at here may be considered a natural outgrowth of the manufacturing problem we analyzed earlier; that is, we are now going to think about the logistics involved in selling some of the products we produced before.

Three furniture stores have ordered desks: Jim's Furniture Emporium wants 30 desks, Dick's Custom Furniture wants 50 desks, and Hugh's Furniture Bazaar wants 45. We have made 70 of our available desks at plant 1, and 80 at plant 2. It will cost $0.50 per desk per mile to ship the desks, and we would obviously like to minimize the shipping costs. A table (or matrix) of the distances between plants and shops is given in Table 9.1. The cost of shipping is easily calculated as

shipping cost to a store
= (cost per desk per mile) · (number of miles) · (number of desks for that store)

What we thus need to determine is the number of desks that go to a particular store from a specified plant.

As we did previously in the manufacturing problem, we may wonder

Table 9.1
Store-to-Plant Distances

	Furniture store		
	(1) Jim's	(2) Dick's	(3) Hugh's
Plant 1	10	5	30
Plant 2	7	20	5

if a solution cannot be obtained by choosing to ship first along the shortest (and cheapest) possible routes. Here, for example, we would start by sending 50 desks from plant 1 to Dick's, 45 desks from plant 2 to Hugh's, and so on; and in this case, such an approach would yield the optimal solution. However, in this problem there are very few plant–store combinations to consider, and there is also adequate supply available at each plant to serve the shortest routes. But it is not difficult to envision problems sufficiently more complicated to make the choices among alternative shipping routes rather difficult. Again, although we have chosen a transportation example that is perhaps too simple, we want to reiterate that the minimization of costs requires the same kinds of trade-offs among alternative choices as does the maximization of the manufacturer's profits.

This transportation problem can be represented as a network problem, as displayed in Figure 9.9. The circles represent *nodes*, which are points where desks are supplied or collected. (In more elaborate problems they may also be points to which material is supplied and from which material is then distributed.) The directed line segments represent *links*, which are all the possible paths along which furniture can be shipped. One possible solution, not necessarily the optimal one, is represented in the partial network of Figure 9.10. We can calculate the cost of this solution, which meets all demand criteria but leaves an

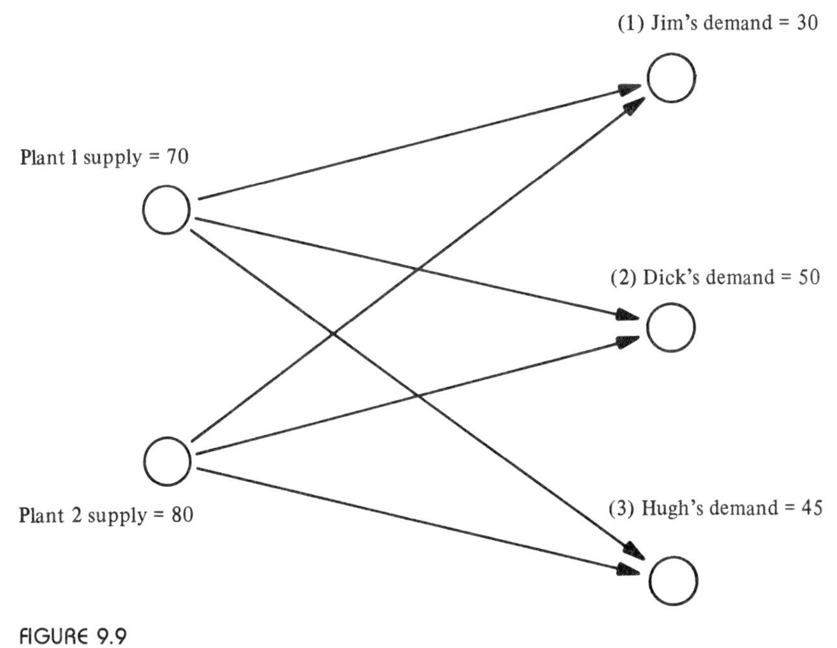

FIGURE 9.9
Network representation of furniture transportation problem

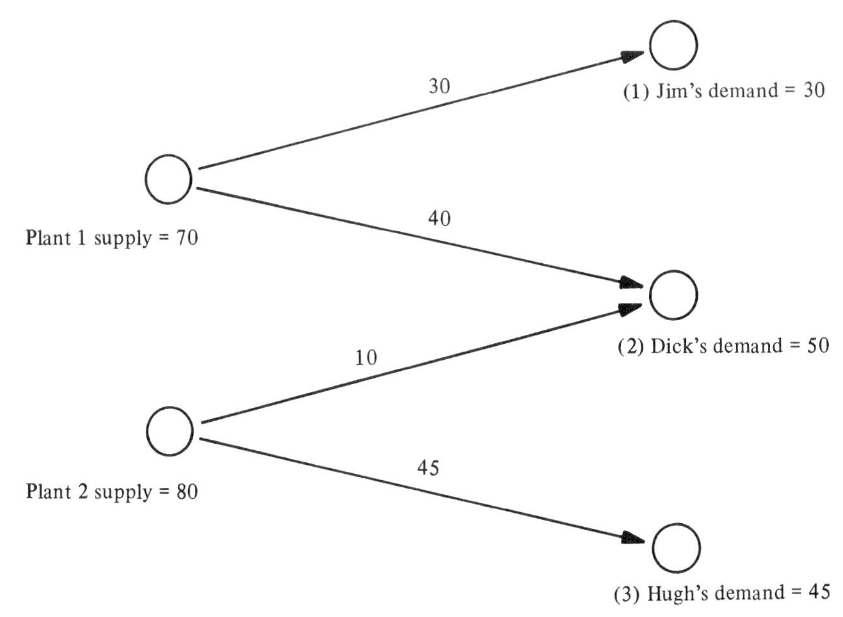

FIGURE 9.10
A possible solution to the furniture transportation problem

excess supply of desks at plant 2, and it turns out to be

$$\text{shipping cost} = (\$0.50)[30 \text{ desks} \cdot 10 \text{ miles} + 40 \text{ desks} \cdot 5 \text{ miles} + 10 \text{ desks} \cdot 20 \text{ miles} + 45 \text{ desks} \cdot 5 \text{ miles}] = \$462.50$$

There are in fact any number of other solutions to this problem, and the solution given here is not the optimal one; that is, it is not the cheapest. The optimal solution has a cost of \$342.50 and is characterized by a distribution of $x_{12} = 50$, $x_{21} = 30$, $x_{23} = 45$; the x_{ij} are defined in the following paragraph.

We have also noted that the transportation problem can be formulated as a linear programming problem. We now turn to this formulation, after introducing some appropriate notation. In Table 9.1 we had assigned to each of the three stores a number exactly for this purpose. We wish to calculate the number of desks to be shipped from plants (1, 2) to stores (1, 2, 3). Let x_{ij} be used to denote the number of desks shipped from plant i to store j. Then, for example,

x_{11} = number of desks from plant 1 to store 1 (Jim's)
x_{12} = number of desks from plant 1 to store 2 (Dick's)
x_{21} = number of desks from plant 2 to store 1 (Jim's)
x_{23} = number of desks from plant 2 to store 3 (Hugh's)

Noting that the cost per mile per desk is a constant (\$0.50), we use the x_{ij} and the distance data in Table 9.1 to calculate a cost (or objective function):

$$\text{cost} = (\$0.50)[10x_{11} + 5x_{12} + 30x_{13} + 7x_{21} + 20x_{22} + 5x_{23}] \quad (9.13)$$

The constraints in this problem arise from the ideas of supply and demand. The two plants cannot produce more desks than their individual capacities, so

$$\begin{aligned} x_{11} + x_{12} + x_{13} &\leq 70 \\ x_{21} + x_{22} + x_{23} &\leq 80 \end{aligned} \quad (9.14)$$

The stores, in turn, must have shipped to them at least enough desks to satisfy their demand:

$$\begin{aligned} x_{11} + x_{21} &\geq 30 \\ x_{12} + x_{22} &\geq 50 \\ x_{13} + x_{23} &\geq 45 \end{aligned} \quad (9.15)$$

Finally, of course, all the desks are real, so that we have a nonnegativity requirement on the numbers of desks that

$$x_{ij} \geq 0 \quad (9.16)$$

The total problem of minimizing the cost function of Eq. (9.13), subject to the constraints of Eqs. (9.14)–(9.16), is clearly a linear programming problem.

There is a restricted version of this linear programming problem that is the classical transportation problem. In this problem, the total supply is set equal to the total demand. As a consequence, all the inequality constraints are simple equality constraints, and then we have in effect one less independent constraint than in the original formulation. We can illustrate how this occurs by returning to our example problem, only now we shall reduce the supply so that plant 1 produces only 45 desks. The total supply is then reduced to 125 desks between the two plants, which is exactly the total demand among the three stores. With equality constraints replacing inequalities, we have on the supply side

$$\begin{aligned} x_{11} + x_{12} + x_{13} &= 45 \\ x_{21} + x_{22} + x_{23} &= 80 \end{aligned} \quad (9.17)$$

while on the demand side we have

$$
\begin{aligned}
x_{11} + x_{21} &= 30 \\
x_{12} + x_{22} &= 50 \\
x_{13} + x_{23} &= 45
\end{aligned}
\tag{9.18}
$$

Now, if Eqs. (9.17) are added to each other, we have a constraint on the total supply; that is,

$$x_{11} + x_{12} + x_{13} + x_{21} + x_{22} + x_{23} = 125 \tag{9.19}$$

By adding Eqs. (9.18), we also obtain a constraint on total demand. It is a simple calculation to show that the total demand constraint equation is identical to Eq. (9.19), the total supply constraint. Hence, because of the equality of supply and demand given by Eq. (9.19), the set of Eqs. (9.17) and (9.18) represents only four—not five—independent equations.

This reduction in the number of independent constraints actually produces some computational benefits in solving classical transportation problems. One of these benefits, incidentally, is that all the variables automatically come out as integers when the constraints are expressed in integers. Even more to the point, however, is the fact that the computation of the optimum proceeds in a comparatively efficient and straightforward way because of the "extra" constraint equation produced by equating demand to supply.

Note that if demand exceeds supply, even the linear programming approach can't get started because it is impossible to get into the feasible region—from which solutions derive. This is clear simply from adding all the supply constraint inequalities, where

$$\sum_{i,j} x_{ij} \leq \text{supply}$$

and comparing it to the sum of the demand inequalities,

$$\sum_{i,j} x_{ij} \geq \text{demand}$$

Of course, if supply exceeds demand, so that there is a positive surplus, a linear programming solution can proceed straightforwardly.

NETWORK ANALYSIS

Another model that we wish to display is that of *network analysis*. To do this we shall expand the transportation network outlined in the preceding section. In that network (see Figure 9.9) each link between nodes was unidirectional, that is, there was flow or movement of goods in only one direction. Further, at each node there was either outflow or inflow; there was no balancing of flow at these interior nodes. In business or economic terms, the previous network did not allow for the return of (unsold) desks, and it did not allow shipping between plants or between stores. In our expanded model we shall show how these activities can be incorporated, as can the return of desks from any store to either plant.

To begin with the inclusion of flow in both directions, we relabel the nodes of our earlier network; we show this relabeling in Figure 9.11. (We shall insert a node 1 later on.) We could assign the same plant-to-store transportation costs as in the previous example, and we can now assign equal or unequal return costs to flow from the stores back to the plants. Costs on the links between the two plants and between the three stores can also be incorporated now. In all of these cases, as we shall see in the mathematical formulation that follows, we can incorporate costs that vary from link to link, that vary in different directions along a link, and that are reflective of real world costs.

The type of network shown in Figure 9.11 is called a *flow network*, and here the flow variable might be the number of desks per unit time. The word flow is evocative of the phenomena associated with a fluid in a piping system, and the evocation is intentional. However, this imagery also suggests that the various links in Figure 9.11 ought to have assigned *bounds*, both lower and upper, on their capacities. This would mean that a given link, in a given direction, must be assigned both a minimum flow and a maximum capacity. The lower bounds are generally zero, although this is not always the case, and the upper bounds may either be infinite or fixed by some realistic circumstance such as shipping capacity.

For conceptual and computational purposes, flow networks are generally drawn so as to include one *source node* (node 1 in Figure 9.12) and one *sink node* (node 7 in Figure 9.12). Note that the links from the source and to the sink are unidirectional. On the source side, since in our example plants 1 and 2 can supply no more than 70 and 80 desks, respectively, the links 1–2 and 1–3 must have maximum capacities of 70 and 80 desks, respectively. Similarly, the three stores have minimum demands of 30, 50, and 45 desks at the respective nodes 4, 5, and 6. Then, the links 4–7, 5–7, and 6–7 must have these figures as representing their minimum flows.

In order to determine a minimum shipping cost or a maximum profit, we can formulate a linear programming problem for the expanded network of Figure 9.12. Let us consider the minimum cost for our schematic presentation. By analogy with Eq. (9.13), we have here an objective

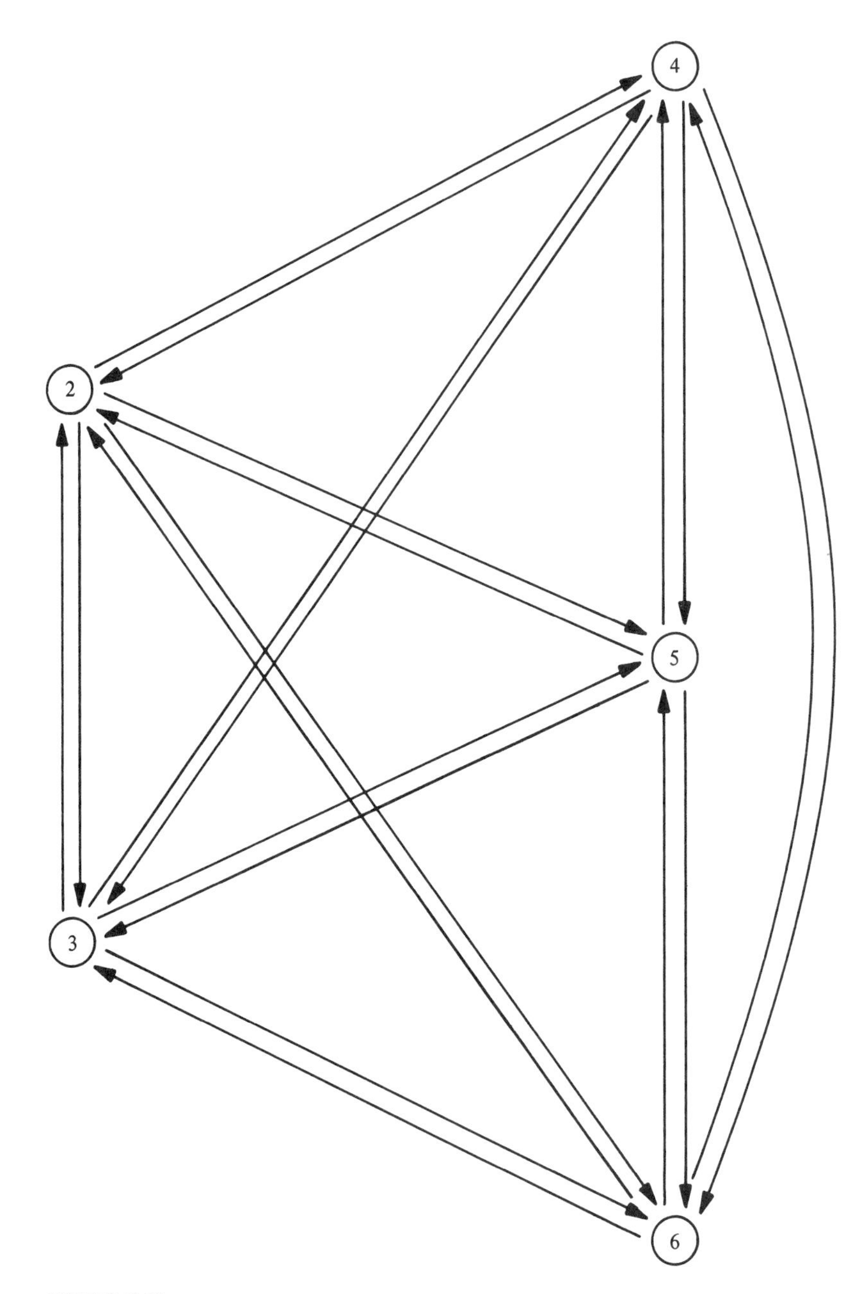

FIGURE 9.11
Expanded transportation network

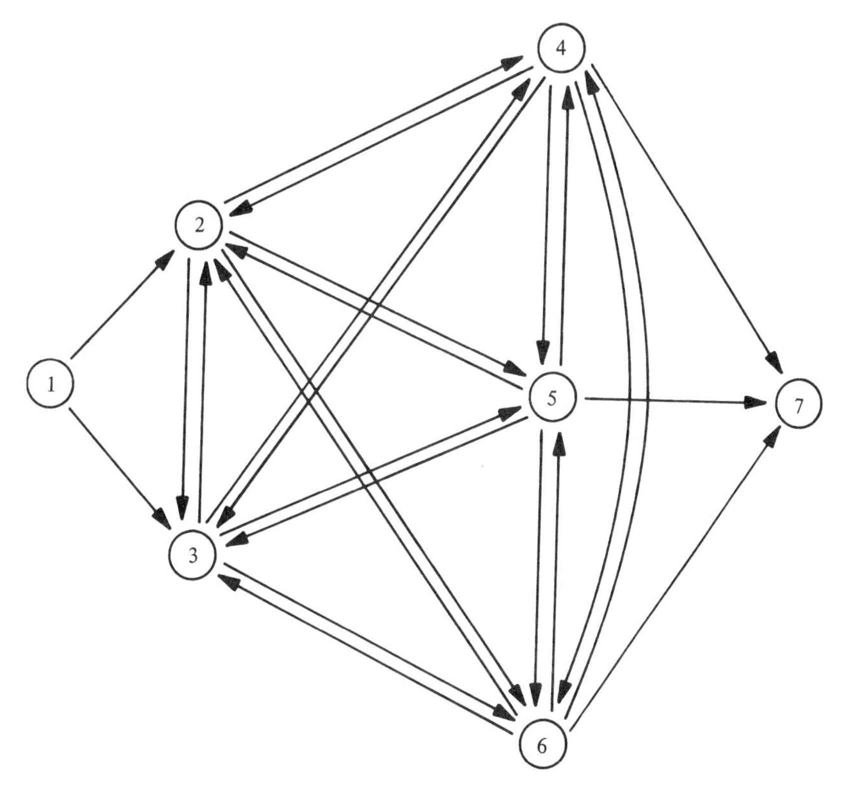

FIGURE 9.12
Flow network with source and sink

function such that we want to

$$\text{minimize} \quad \sum_{i=1}^{7} \sum_{j=1}^{7} c_{ij} x_{ij} \tag{9.20}$$

where the c_{ij} represent the net costs in each of the links, and x_{ij} now represents the flow in the link from node i to node j. If there is no direct link between these nodes, we could either stipulate that a particular x_{ij} (or its upper bound) is zero, or that the cost of shipping between those nodes is infinite; that is, $c_{ij} = \infty$. These net costs could be calculated so as to include benefits (in the dimension of money) from choosing a particular link, as well as the actual or perceived shipping costs for the various links. Note that there is no requirement that the cost of shipping from node i to node j be equal to the cost of shipping from node j to node i. Hence, in general, $c_{ij} \neq c_{ji}$.

The constraint equations derive from four considerations: balance of flow at a node, upper bound constraints, lower bound constraints, and nonnegativity. The flow balance constraints are analogous to the equality versions of Eqs. (9.14) and (9.15), as well as to Kirchhoff's laws as used in Chapter 6. The basic rule is that "what goes in must come out, excepting any local supply or consumption." In symbolic terms, if s_j is used to denote the flow generated or goods created at the jth node, and d_j the local demand or consumption at that node, then

$$s_j + \sum_{\substack{i=1 \\ i \neq j}}^{7} x_{ij} = d_j + \sum_{\substack{k=1 \\ k \neq j}}^{7} x_{jk}, \qquad j = 1, 2, \ldots, 7 \tag{9.21}$$

For the source node, for example, since there are direct links only to nodes 2, 3, we would write

$$s_1 = x_{12} + x_{13} \tag{9.22}$$

while for the sink node,

$$x_{47} + x_{57} + x_{67} = d_7 \tag{9.23}$$

The remaining sets of constraints are straightforwardly written out. The maximum capacities in each link are represented by upper bounds U_{ij}, so that

$$x_{ij} \leq U_{ij} \qquad \text{for all} \quad i, j = 1, 2, \ldots, 7 \tag{9.24}$$

Similarly, the minimum capacities are expressed as lower bounds L_{ij}, so that

$$x_{ij} \geq L_{ij} \qquad \text{for all} \quad i, j = 1, 2, \ldots, 7 \tag{9.25}$$

Finally, of course, all the flow variables are positive,

$$x_{ij} \geq 0 \qquad \text{for all} \quad i, j = 1, 2, \ldots, 7 \tag{9.26}$$

This linear programming formulation [Eqs. (9.20), (9.21), (9.24)–(9.26)] is typical of a large group of network problems for which there exist rapid and efficient solution procedures. We note that, in contrast to our earlier network analysis, it is possible to use the c_{ij} elements of the objective function to include explicitly the benefits of shipping as well as the costs of shipping. Also, we can guess that to the extent that there are positive lower bounds L_{ij} that are greater than the zeroes of the corresponding nonnegativity conditions, the lower bounds provide the driving force for the optimization process. Without the lower bounds, there is nothing to force the flow of goods (or whatever) between links. However, in the absence of lower bounds, the inclusion of benefits of shipping or flow could provide the driving force.

We should note, too, that network analysis has applications other than to minimizing a transportation cost. One application is to *location analysis*, where the object is to determine the optimal location of some service outlets. For example, where should firehouses be located in a city street network? The answers would depend on local housing density, street access, and so on. The search for an optimal answer could be modeled with a network analysis. *Routing* problems can also be modeled with a network, enabling us to plan bus routes, garbage truck routes, or routes for traveling salespeople. We also point out that network problems can be phrased in other terms, for example, using minimum shipping times or shortest paths between links as the objective functions. All in all, network analysis has many interesting applications to problems that touch our daily lives.

SUMMARY

This chapter has been devoted to one of the more important aspects of operations research, that of optimization in contexts where the objective function can be written as a linear function of the variables of interest. This is the collection of optimization problems that can be solved with linear programming techniques. We have seen that a variety of practical problems can be profitably modeled in this way, some of these problems being the product-mix problem, the transportation cost problem, and general network problems. In all of these cases the emphasis has been on recognizing and setting up the correct linear programming problem. We have not developed any solution techniques (except for the graphical solution to the simple product-mix problem) because these techniques are not appropriate for this level of discussion. The important point here is the recognition that many such optimization problems can be treated as linear programming problems.

We should also note that as important as linear programming is to operations research, it is only one part of it. In addition to the optimization techniques of nonlinear, dynamic, and integer programming, operations research comprises many other interesting approaches, including queueing theory, game theory, and simulation (particularly Monte Carlo simulation). These approaches are concerned with such widely varying problems as assessing the costs, to the consumer and to the provider of service, of having too few or too many service lines in a service facility; with trying to rationalize economic and strategic decisions in the face of uncertainty; and with developing simulation techniques for solving problems that are too intractable to be solved analytically and too expensive to be treated experimentally. Thus, while we have seen some interesting formulations of problems in this chapter, we have by no means exhausted the field of operations research.

Problems

9.1 (a) Find the extreme values of the function $y = \sin x$ for $0 \leq x \leq \pi$. Are the extreme values maxima or minima?
(b) What are the maxima and minima of $y = \sin x$ in the interval $0 \leq x \leq 2\pi$?

9.2 (a) What are the extreme values of $y = x$ in the interval $0 \leq x \leq 2\pi$?
(b) What are the extreme values of $y = x - x^3/3!$ in the interval $0 \leq x \leq 2\pi$?
(c) How do the answers to this question relate to the answers to Problem 9.1?

9.3 A string of length l can be used to outline many simple geometrical figures, such as an equilateral triangle with sides $l/3$, a square with sides $l/4$, a pentagon with sides $l/5$, or a circle of circumference l. Calculate the areas of several figures and show that the area increases with the number of sides. What would then be a good intuitive guess as to the maximum area that can be enclosed by a string of given length?

9.4 Find the maximum area of a triangle that can be inscribed in a semicircle of diameter d. Do this by showing that the area of a triangle is $bc/2$ and that the variable c can be eliminated through a relation that can be derived between the sides of the triangle and the diameter of the circle.

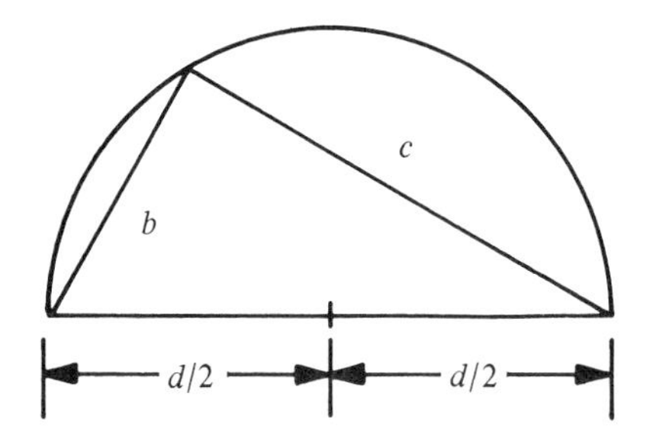

9.5 Solve, graphically, the following linear programming problem involving nonnegative variables x_1 and x_2.

Maximize $z = 5x_1 + 3x_2$

subject to

$$3x_1 + 5x_2 \leq 15$$
$$5x_1 + 2x_2 \leq 10$$

9.6 Solve, graphically, the following linear programming problem.

$$\text{Maximize} \quad z = 2x_1 + x_2$$

subject to

$$4x_1 + 3x_2 \leq 24$$
$$3x_1 + 5x_2 \leq 15$$
$$x_1, x_2 \geq 0$$

9.7 Solve, graphically, the following linear programming problem.

$$\text{Maximize} \quad z = 2x_1 + x_2$$

subject to

$$x_1 + x_2 \leq 4$$
$$3x_1 + x_2 \leq 10$$
$$x_1, x_2 \geq 0$$

9.8 Formulate the following product-mix problem as a linear programming problem. Assume that a certain manufacturing concern regularly produces three different products, which are sold for \$6, \$11, and \$22 per unit, respectively. Further, assume that these prices do not vary with the firm's output, that is, the market can absorb any amount of the products it may turn out without adverse effects on their prices. A total of four input factors is required for their production, and the amounts of each input factor required for each unit of the three products, together with the unit costs and the available supplies of these inputs, are given in the accompanying table. Assuming no other restrictions on the firm's production decisions, how much of each product should be produced in order to maximize profit?

Input	Unit cost ($)	Product 1	2	3	Supply of input
1	2	0	1	2	150
2	1	1	2	1	200
3	0.5	4	6	10	400
4	2	0	0	2	100

9.9 A vendor customarily mixes three types of washers in different proportions for sale at different prices. The specifications of two washer mixtures and the available supply of ingredients are given in the accompanying table. The costs of brass, steel, and aluminum washers are $0.60, $1.20, and $1.00 per pound, respectively. The mixtures are sold for $1.50 per pound of A and $1.80 per pound of B. The vendor wishes to know how much of A and B should be prepared. Formulate this as a linear programming problem.

Mixture	Brass	Steel	Aluminum
A	$\frac{1}{4}$	$\frac{1}{2}$	$\frac{1}{4}$
B	0	$\frac{1}{2}$	$\frac{1}{2}$
Supply (lb)	1000	400	400

9.10 A trucking firm has received an order to move 3000 tons of miscellaneous goods. The firm has available at the moment a fleet of 150 trucks of 15-ton capacity and another fleet of 100 trucks of 10-ton capacity. The operating costs of these trucks are $30.00 and $40.00 per ton, respectively. The firm also has a policy of retaining at least one 15-ton truck with every two 10-ton trucks in reserve. How many of the two classes of vehicles should be dispatched to move the goods at minimal operating costs? (Formulate this as a linear programming problem.)

9.11 Two ingredients are mixed in varying proportions to make two kinds of oil, called massage oil and machine oil, which are both sold at wholesale for $3.00 per quart. The cost of massage oil is $1.50 per quart, while that of machine oil is $2.00 per quart. Although there is no strict formula to follow in mixing the ingredients, two rules are generally observed: (1) Massage oil may contain no less than 25% of the first ingredient and no less than 50% of the second; and (2) machine oil may contain no more than 75% of the first ingredient. There are, respectively, 30 quarts and 20 quarts of the two ingredients available. The object is to determine the most profitable quantities of the two oils. Formulate this as a linear programming problem.

10

DIFFRACTION AND SCALE

Diffraction is a phenomenon that occurs when waves of any type strike an object. Typical waves may be sound, water, or light waves. The result of a wave striking an object is the bending of the wave around that object. This enables us, for example, to hear sound around corners. When waves pass through an opening, they bend around the edges of the opening in such a way that a *diffraction pattern* is formed. For visible light passing through a slit the diffraction pattern consists of alternating light and dark regions. Such a diffraction pattern can be seen by looking at a distant light source through a narrow crack like that which might be formed between a couple of extended fingers. We shall be concerned here with seeing what happens to *radiation* from a source, and investigating how radiated waves interact with each other and with their surroundings.

Diffraction patterns are highly dependent on the size of the object in the path of the waves in relation to the distance between wave peaks, that is, the *wavelength* of the incoming wave. (The wavelength λ of the wave is equal to the product of the speed c of the wave and the period T: $\lambda = cT = c/f = 2\pi c/\omega$. Here, again, f is frequency and ω is circular frequency.) Because of the dependence of diffraction patterns on relative sizes, it is clear that *scale effects* will be very important in our discussion of diffraction. We shall see the importance of scaling in the development of the geometry of diffraction and in the analysis of diffraction gratings. Then we shall discuss examples of diffraction, including wheels going over depressions (in the discussion on geometry), x rays in crystal lattices, and sound waves over barriers. Also, we shall comment throughout on how diffraction phenomena are useful in various research endeavors.

We shall impose one restriction in our discussion of diffraction: We shall consider only *planar waves*, this class of problems being called *Fraunhofer diffraction*. Although many sources emit waves that are other than planar, such waves can be treated as planar at sufficiently large distances from the source (again, a question of scale). The more general problem of Fresnel diffraction, which treats nonplanar wave diffraction, will not be discussed here.

DIFFRACTION GEOMETRY

Diffraction patterns result from the interaction of several oscillating wave sets, so we shall first discuss the concepts of superposition and of interference. The *principle of superposition* states that when two or more waves are passing through the same region in space at the same time, the net effect is that which could be calculated from the addition, or superposition, of the waves at every instant of time and at every point in space. Let us represent three sets of waves as shown in Figure 10.1a, 10.1b, and 10.1c. Note that displacement x above the time t axes is considered positive, while displacement below the time axes is negative. The superposition of the first two waves in each case illustrated in Figure 10.1 is then simply the addition of the amplitudes of the waves 1 and 2 at every instant of time.

Whenever two or more waves merge, the resultant wave represents the *interference* of the original waves. We say that *constructive interference* has occurred if the amplitude of the resultant wave is greater than that of either of the two original waves. However, if the resultant wave has a smaller amplitude than either of the original waves, *destructive interference* has occurred. If all of the incoming waves are of the same wavelength λ (as in Figure 10.1) the radiation is said to be *monochromatic*, and the resulting interference will yield *monochromatic diffraction*. Then two such waves, with their crests lined up in the same place at the same time, undergo constructive interference to produce a wave with twice the amplitude of the original waves. (See Figure 10.1a.) These waves are said to be *in phase*. Conversely, the first two waves of Figure 10.1b sum to produce a straight line along $x = 0$. This is a case of complete destructive interference. Here the waves are said to be completely *out of phase*. For waves of visible light, constructive interference produces bright spots and destructive interference produces dark spots.

Consider the case of a distant monochromatic light source illuminating a double slit, a double slit being nothing more than a pair of gaps in an otherwise opaque barrier (as in Figure 10.2). We wish to examine the brightness at some point P on a screen located far enough away that ray lines r_1 and r_2 become essentially parallel ($D >> d$), that is, so the

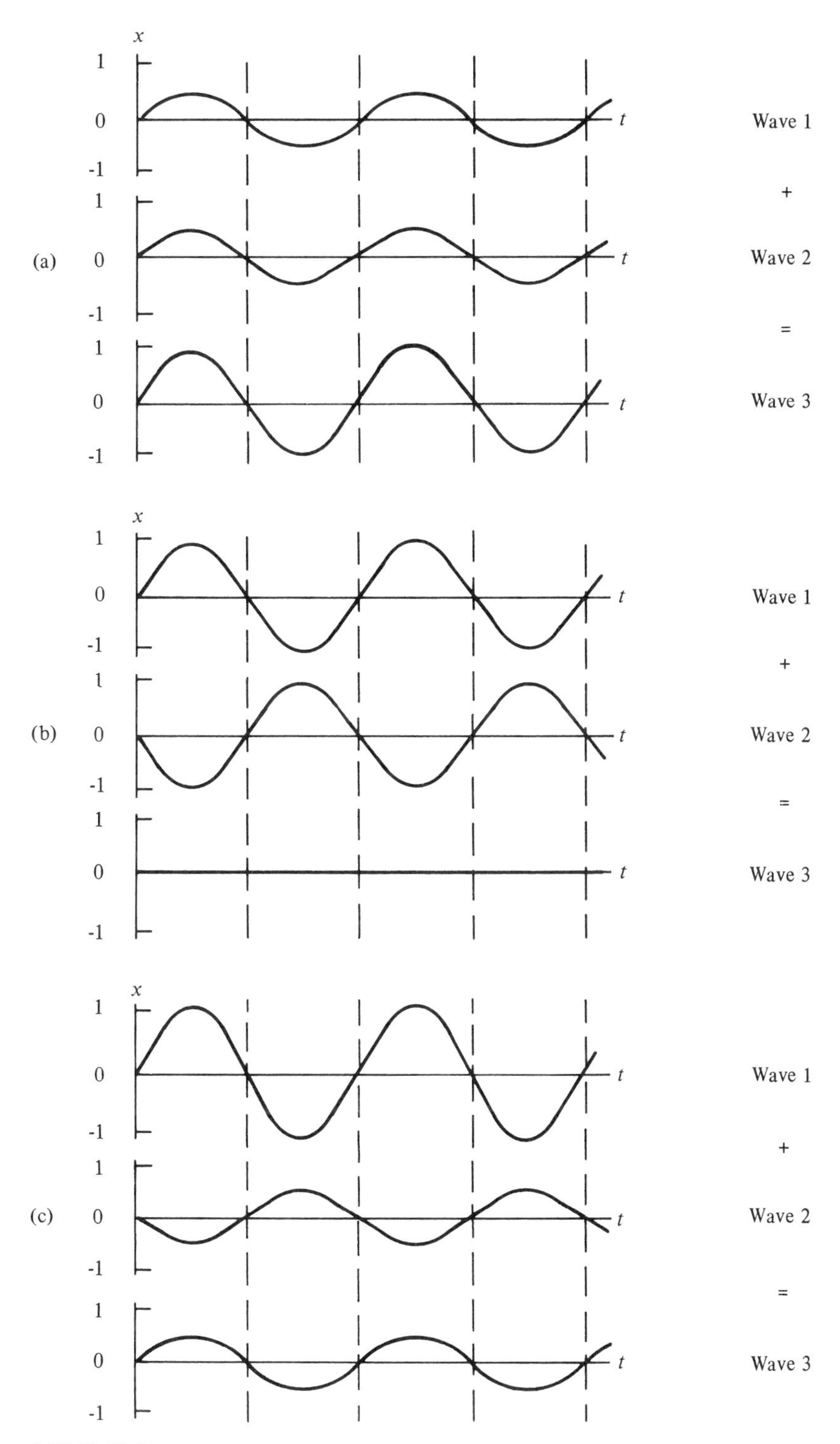

FIGURE 10.1
The superposition of waves 1 and 2 to form wave 3

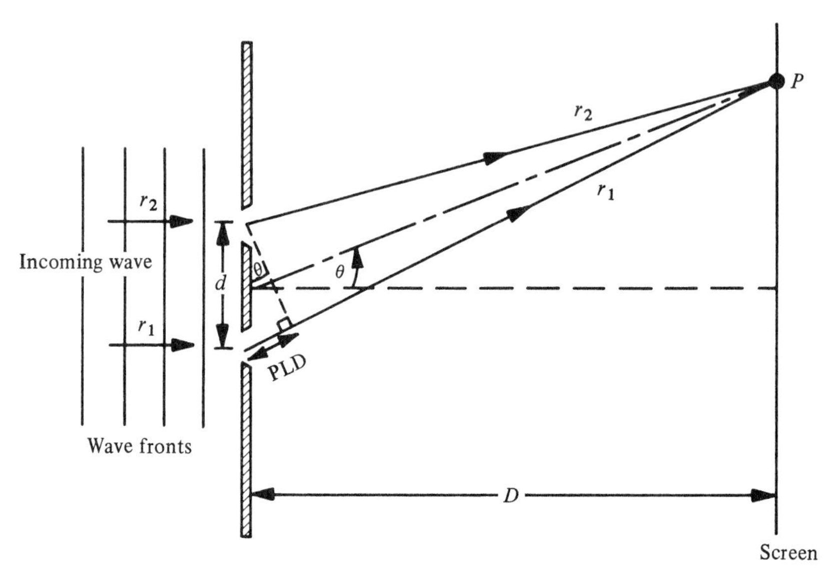

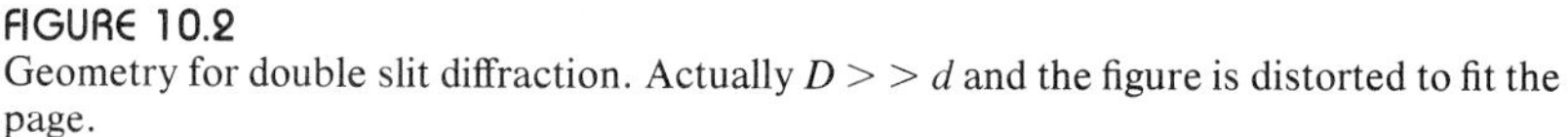

FIGURE 10.2
Geometry for double slit diffraction. Actually $D >> d$ and the figure is distorted to fit the page.

incoming radiation is again planar, being so represented here by parallel line wave fronts parallel to the screen. We now look at the differences in ray path lengths r_1 and r_2. (The horizontal scale in Figure 10.2 is shortened for clarity, and so that it will fit on the page!) When the difference in path length is an exact whole-number multiple of the wavelength λ, we will see that the two waves merge at point P in phase (constructive interference) and we have a bright spot, a point of maximum amplitude.

The path length difference PLD is shown on Figure 10.2. It can be calculated from

$$\text{PLD} = d \sin \theta$$

For a bright spot to occur, the path length difference must contain an integral number of wavelengths, or

$$\text{PLD} = m\lambda = d \sin \theta, \qquad m = 0, 1, 2, \ldots \tag{10.1}$$

Bright spots (maxima) will occur for different values of θ as the integer m takes on different values. There is a central maximum at $\theta = 0°$ described by $m = 0$, and there are successive maxima on either side of the central bright area described by the *order* m (see Figure 10.3).

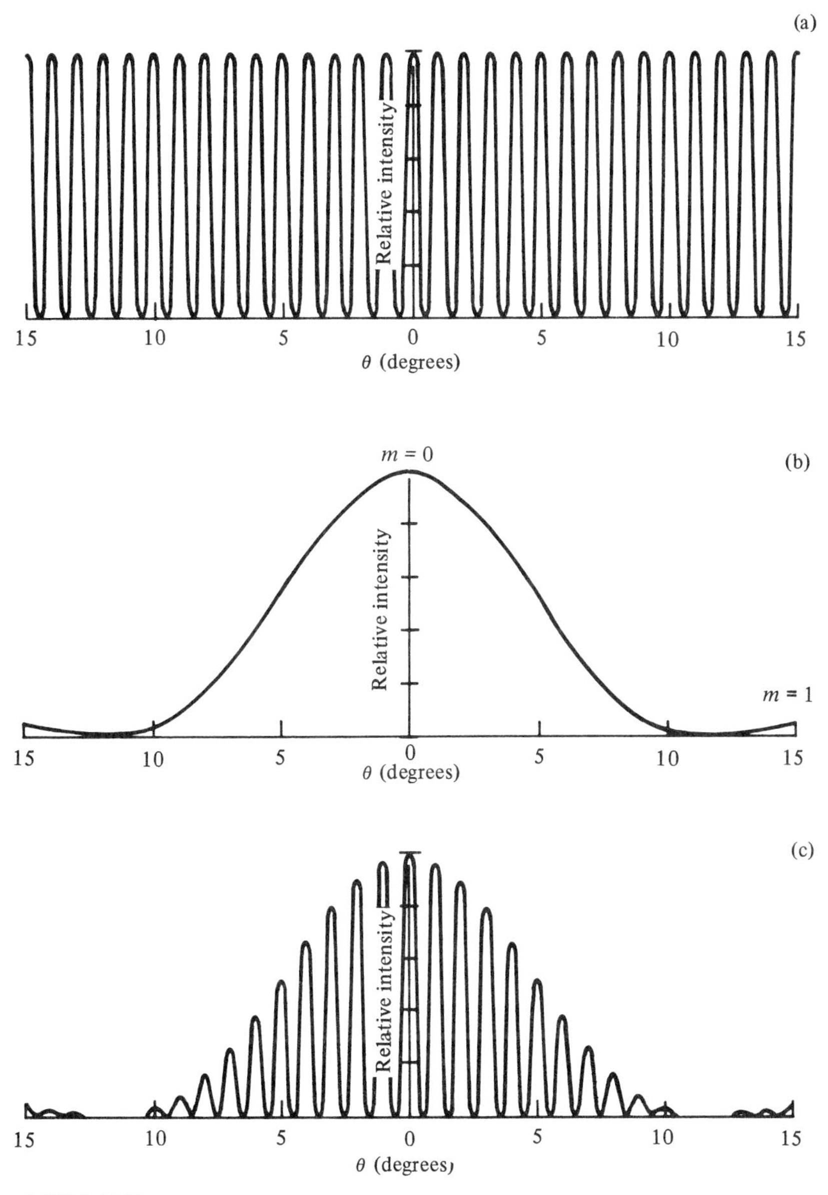

FIGURE 10.3
(a) Interference pattern due to double slit interference only; (b) single slit diffraction pattern; (c) true double slit interference pattern, a combination of (a) and (b). Here the slit width is $a = 5\lambda$. (Reprinted from D. Halliday and R. Resnick, *Fundamentals of Physics*. Copyright © 1970. Reprinted by permission of John Wiley and Sons, Inc.)

Interspersed between the maxima are minima (dark spots), which occur when the PLD is equal to $\frac{1}{2}\lambda$ or some odd multiple thereof. Being one half of a wavelength out of phase results in complete destructive interference (see Figure 10.1b). Therefore, for minima to occur

$$\text{PLD} = \left(m + \frac{1}{2}\right)\lambda = d \sin\theta, \qquad m = 0, 1, 2, \ldots \tag{10.2}$$

The double slit diffraction pattern is really a combination of two phenomena, each providing alternating dark and light patterns. Light passing through a single slit bends to form a diffraction pattern by combining rays from different parts of the same slit in the same way we superposed rays r_1 and r_2 in Figure 10.2. (This pattern is what we observe when looking through our fingers.) When more than one slit is used, the interference pattern formed by rays coming from each slit separately superimposes itself upon the diffraction pattern produced by a single slit alone. This is illustrated graphically in Figure 10.3 and photographically in Figure 10.4.

The equations governing single slit diffraction are similar to Eqs. (10.1) and (10.2) except for two notable variations. The distance d between slits is replaced by the width a of the slit, and the minima appear to occur under maximum conditions (and vice versa) since

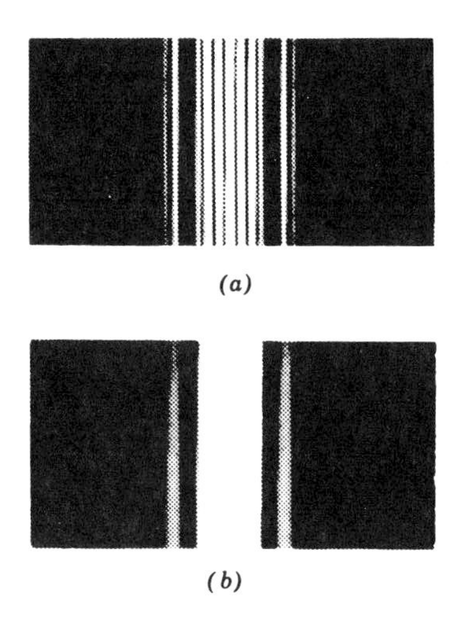

FIGURE 10.4
(a) Single slit diffraction pattern; (b) double slit pattern showing interference fringes. (Reprinted from D. Halliday and R. Resnick, *Fundamentals of Physics*. Copyright © 1976. Reprinted by permission of John Wiley and Sons, Inc.)

$$m\lambda = a \sin\theta, \qquad m = 1, 2, 3, \ldots \quad \text{(minima)} \qquad (10.3)$$

and

$$\left(m + \frac{1}{2}\right)\lambda = a \sin\theta, \qquad m = 1, 2, 3, \ldots \quad \text{(maxima)} \qquad (10.4)$$

However, this apparent reversal of maxima and minima is not real. It arises from the geometry of single slit diffraction. Suffice it to say here that a path length difference of $\frac{1}{2}\lambda$ between two rays diffracting through a single slit still produces a dark spot; that is, total destructive interference.

We see from Eqs. (10.1)–(10.4) that the position θ of the fringes (alternating light and dark areas) depends on the three variables λ, a, and d. In particular, the relationship of wavelength λ to slit width a, and the relationship of λ to the distance between the slits d, will be discussed in detail in the section on diffraction gratings. First, however, we will examine a vibration problem that is analogous to those interference phenomena. The purpose of this diversion is to illustrate similarities between two branches of science.

Consider the wheel of a railroad car passing over a low spot on the rail (as shown in Figure 10.5). As the wheel traverses the low spot, the vertical displacement of the wheel is a combination of the variable depth of the low spot and the additional deflection of the rail due to the

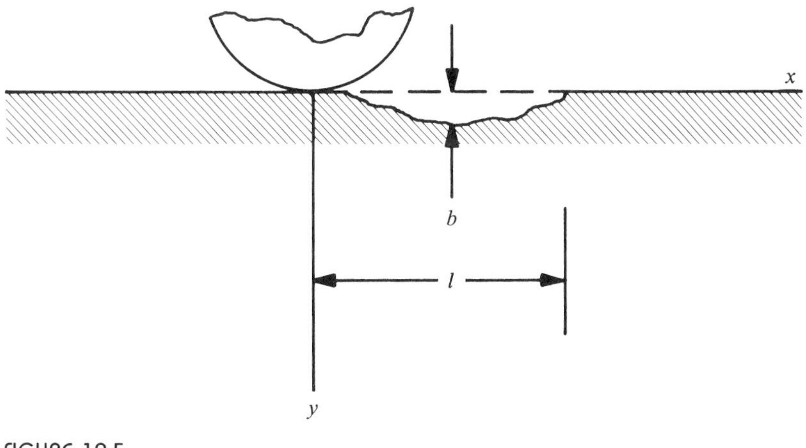

FIGURE 10.5
Wheel and low spot on rail

weight of the wheel interacting with the low spot. If we let u represent the variable depth of the low spot [$u = f(x)$] and y the additional vertical deflection of the rail as the wheel passes by, we can write the equation of motion of the wheel in the vertical direction as

$$m \frac{d^2(y + u)}{dt^2} + Ky = 0 \tag{10.5}$$

Here m is the mass of the wheel and K is the vertical load that produces a deflection of unity. Equation (10.5) is of the same form as the equation for the motion of a spring–mass system (Chapter 6).

With the aid of the chain rule, the term d^2u/dt^2 can be written as a function of the speed v of the train. When this substitution is made, Eq. (10.5) becomes

$$m\left(\frac{d^2y}{dt^2} + v^2 \frac{d^2u}{dx^2}\right) + Ky = 0$$

or

$$m \frac{d^2y}{dt^2} + Ky = -mv^2 \frac{d^2u}{dx^2} \tag{10.6}$$

Now if we define the shape of the low spot by

$$u(x) = \frac{b}{2}\left(1 + \cos \frac{2\pi x}{l}\right) \tag{10.7}$$

with l representing the length of the low spot (see Figure 10.5), and b the depth of that spot at its midpoint, then we can find d^2u/dx^2 and substitute it into Eq. (10.6). Remembering that $x = vt$ allows us to rewrite Eq. (10.6) as

$$m \frac{d^2y}{dt^2} + Ky = +mv^2 \frac{b}{2} \frac{4\pi^2}{l^2} \cos \frac{2\pi vt}{l} \tag{10.8}$$

and we once again have an equation for a system undergoing forced vibrations, that is, the track is forced to vibrate by the wheel passing over the low spot. The solution of the differential equation (10.8) can be written once we decide on appropriate initial conditions. Since we are looking at the additional deflection of the rail due only to the wheel's passing over the low spot, we will set initial conditions of $y = 0$ and $dy/dt = 0$ when $t = 0$. With the notation that T is the period of vibration of the wheel on the rail, we can write the solution of Eq. (10.8) as

$$y = -\frac{b/2}{1 - \left(\frac{l}{vT}\right)^2}\left(\cos\frac{2\pi vt}{l} - \cos\frac{2\pi t}{T}\right) \tag{10.9}$$

Equation (10.9) tells us that the additional deflection y is directly proportional to the depth of the spot b, and it depends on the ratio l/vT. Further, the variation in the deflection is periodic because of the cosine terms. The period of vibration of the wheel on the rail T is related to the radius of the wheel. Thus, the additional deflection depends primarily on the interaction between the size (radius) of the wheel and the dimensions l and b of the low spot. The deflection increases with the ratio of wheel size to length of the low spot. We noted previously that the deflection also increases with additional depth of the spot. We could say, then, that the additional deflection depends on what the wheel "sees" as it approaches a low spot on the rail.

In a similar fashion interference patterns depend on what the incoming wave "sees" in terms of the dimensions of the slit or slits. Equations (10.1)–(10.4) show the relationships between the wavelength λ of the incoming wave, and the dimensions of the slit width a and the distance between slits d. These distances determine the placement angle θ of the fringes in the resulting interference pattern. Further, the intensity (or square of wave amplitude) I_θ of any of the maxima also depends on the ratios of d/λ and a/λ, since

$$I_\theta = I_{\text{max}}(\cos\beta)^2\left(\frac{\sin\alpha}{\alpha}\right)^2 \tag{10.10}$$

where

$$\beta = \frac{\pi d}{\lambda}\sin\theta \tag{10.11}$$

and

$$\alpha = \frac{\pi a}{\lambda}\sin\theta \tag{10.12}$$

Thus the relative sizes of the interacting wave and the slits determine the pattern of fringes in much the same way that the relative sizes of the wheel and the low spot on the rail determine the additional deflection of the rail. Now we will examine these dimensional interactions in more detail as we discuss the diffraction grating.

DIFFRACTION GRATINGS

An extension of the double slit interference phenomenon can be accomplished by increasing the number of slits N. An object having any number of parallel equidistant slits of the same width is called a *diffraction grating*. In practice, diffraction gratings consist of some planar material (for example, a piece of transparent glass) that has carefully ruled parallel "scratches" on its surface, with the scratches precisely arranged so as to be equally spaced and of equal widths. The scratches serve the same purpose as slits. There are often several hundred scratches to a millimeter!

Increasing the number of slits of constant width a without changing d and λ results in sharper maxima (see Figure 10.6). Another change, of

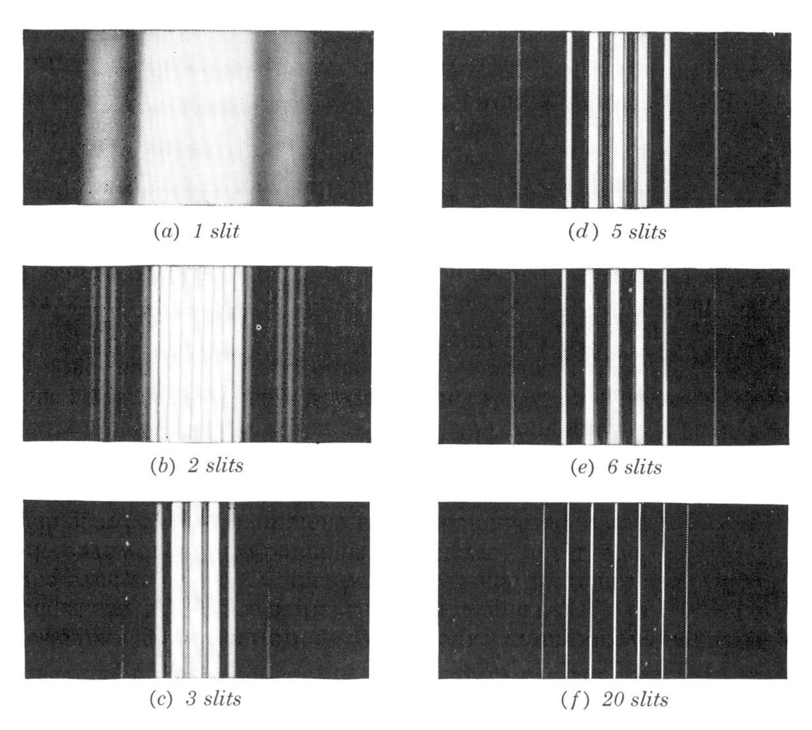

FIGURE 10.6
Diffraction pattern for gratings with an increasing number of slits for the same value of d and λ. (Reprinted from F. A. Jenkins and H. E. White, *Fundamentals of Optics*. Copyright © 1950. Used by permission of McGraw-Hill Book Company.)

less importance, is the appearance of weak *secondary maxima* between the *principal maxima*. This is seen most clearly in Figures 10.6c, 10.6d, and 10.6e. The number of secondary maxima increases with the number of slits N, although the intensity of each secondary maximum becomes weaker and weaker. The secondary maxima are not important in actual gratings, so we will focus solely on the principal maxima.

We note from Figure 10.6 that the positions of the principal maxima do not change with increasing N. This stability is explained by the same argument as that used in the double slit case. Principal maxima will occur when the path length difference between rays from adjacent slits (pairs) is given by

$$d \sin \theta = m\lambda \qquad m = 0, 1, 2, \ldots \tag{10.13}$$

where d is the spacing between slits. Thus the *locations* (or angular separations) of the principal maxima are determined only by the ratio λ/d. This means that if we want to "spread out" an interference pattern, we would have to use a grating whose slits are very close together (see Figure 10.7). It also means that for a given slit spacing d, light of different wavelengths will have principal maxima at different angles θ. Looking at a grating illuminated by white light, we would see a rainbow

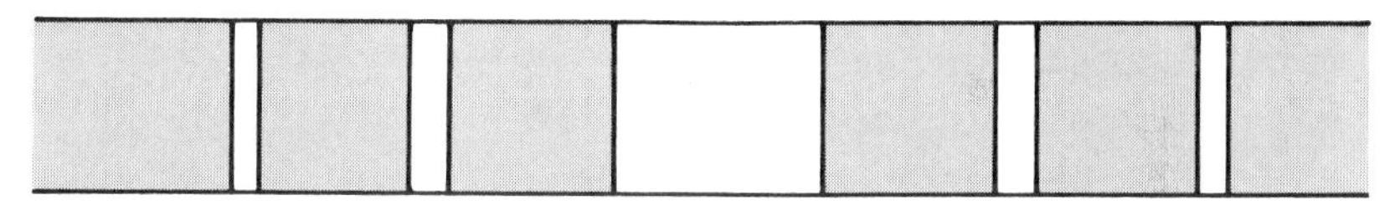

$d \cong 1000\lambda$

$d \cong 100\lambda$

FIGURE 10.7
The spreading out of the principal maxima with decreasing d

of colors. Each wavelength present in white light would have its principal maximum at a different angle from the bright (white) central image ($m = 0$), and therefore each color would be separated from the next, much like what is observed when a prism separates white light into its constituent colors. Since the angle θ can be measured, and since the slit spacing is noted on any grating, the wavelength of any principal maximum can be found by using Eq. (10.13); and gratings are often used to do just that—measure wavelengths very precisely.

Let us now examine the effect of slit width a on the interference pattern. In Eq. (10.10) the factor $(\sin \alpha/\alpha)^2$ gives the intensity distribution due to the diffraction by each single slit. Equation (10.12) shows in turn that α depends on the ratio a/λ. Therefore, the relative intensities of the principal maxima depend on the relationship of the wavelength to the slit width. Figures 10.8 and 10.9 illustrate this dependency for a single slit. In Figure 10.8 we see how the width of the central maximum decreases as the slit width increases. Figure 10.9 displays a photograph of monochromatic light after it has been diffracted by a single slit and allowed to fall on a distant screen.

Note how the maxima to the right and left of the central maximum get progressively weaker. The relative intensities of the principal maxima clearly vary (in Figure 10.9) with changing a/λ ratio. In addition, as the slit width a is narrowed, for a given wavelength λ, the diffraction pattern spreads out. However, this spreading is negligible compared to that introduced by decreasing the distance between slits in a grating.

So far we have shown how changes in the scaling ratios a/λ and d/λ produce changes in interference patterns. At this point we shall look more closely at the relationship of a, d, and λ. Are there any limits imposed on a or d in relation to the wavelength of incident radiation? If the width of the slit is large compared to the wavelength, there is very little diffraction (see Figure 10.9). When the slit width becomes much smaller than a wavelength, diffraction becomes very pronounced and, if there are several slits, the screen is uniformly illuminated. The condition that $a << \lambda$ can rarely be met, so interference fringes are not usually of uniform intensity. Rather, the intensity of the fringes drops off with increasing order m.

Equation (10.13) tells us that as soon as the slit spacing d becomes less than the wavelength, the only solution possible is for $m = 0$. For $m > 0$, Eq. (10.13) would require $\sin \theta > 1$ when $d < \lambda$. This is of course impossible. Thus, if the spacing is too small, there is only one possible maximum, the zero-order (central) maximum. This means, in effect, that higher-order maxima can only be produced when $d > \lambda$.

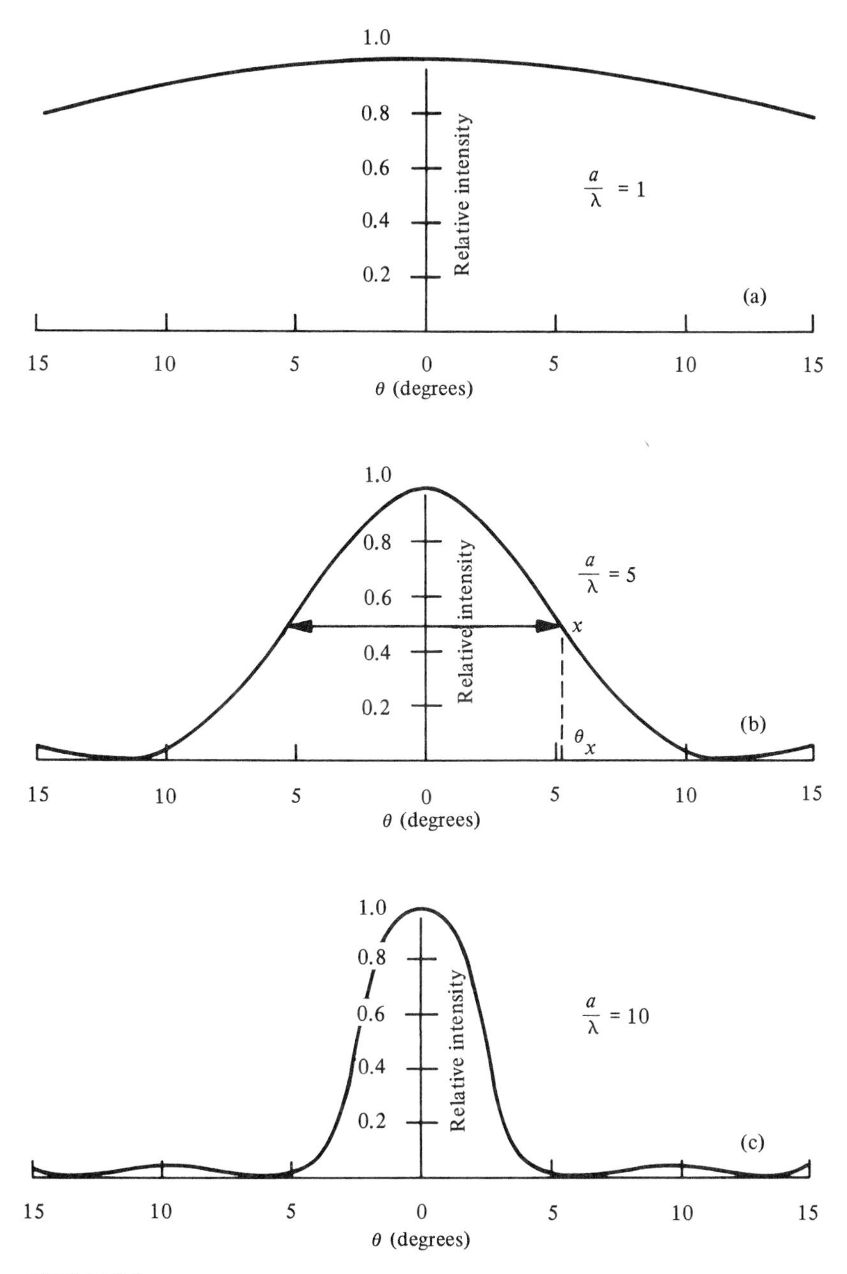

FIGURE 10.8
The relative intensity in single slit diffraction for three values of the ratio a/λ. (Reprinted from D. Halliday and R. Resnick, *Fundamentals of Physics*. Copyright © 1970. Reprinted by permission of John Wiley and Sons, Inc.)

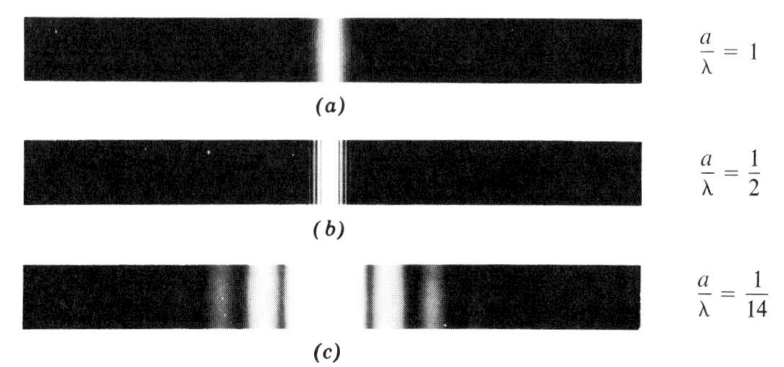

FIGURE 10.9
Photographs of monochromatic light after single slit diffraction through slits of varying widths. (Reprinted from D. Halliday and R. Resnick, *Fundamentals of Physics*. Copyright © 1976. Reprinted by permission of John Wiley and Sons, Inc.)

X-RAY AND ATOMIC PARTICLE DIFFRACTION

Just as gratings can be used to identify and separate incident visible radiation into its component wavelengths, atoms can be used to study x-ray spectra.* Conversely, x-ray diffraction can be used to study the arrangement of atoms in crystals.

X rays were discovered in 1895 by W. Roentgen, who was never able to observe interference with these so-called x rays. However, in 1899 a slight broadening of an x-ray beam was observed after the beam had passed through slits a few thousandths of a millimeter wide; the broadening was assumed to be due to diffraction. Measurements and calculations estimated the wavelength to be about 1 Å.† It remained for Max von Laue to suggest (in 1912) that if x rays were indeed waves, and if they were allowed to propagate in crystals whose atomic spacings were about the same as the wavelengths of the x rays, then an interference pattern would occur. Two important assumptions were confirmed by a further experiment: x rays are a form of electromagnetic radiation with wavelengths having the dimensions of the atomic spacings in crystals; and solids contain atoms arranged in a regular array. Thus, von Laue used his knowledge of diffraction by gratings and the importance of the scaling ratio d/λ to advance scientific knowledge by means of analogy.

Sir William L. Bragg proposed a simple way to analyze the diffraction of x rays by crystals. He proved that atoms lying on any one plane within a crystal act like a partially silvered mirror with respect to an

*Any radiation consists of one or more wavelengths. The delineation of these wavelengths represents the spectrum of the radiation.

†Å represents the unit of length called an angstrom: 1 Å = 10^{-10} m = 10^{-8} cm.

incident wave; that is, they reflect part of the wave and allow the rest to pass through. Knowing this, we can simply deal with the interference between waves reflected from parallel (Bragg) planes rather than with the interference of waves coming from all the atoms individually. Within any crystal structure there are several sets of Bragg planes, as shown in Figure 10.10. Note that each set of planes has its own unique spacing d between successive planes.

Consider a crystal with a series of planes of atoms and an incident beam having a wavelength of the order of the planar spacing d (Figure 10.11). The reflected rays will constructively interfere at some distant point if their path length difference (PLD) is an integral multiple n of the wavelength. From Figure 10.11 we find that the PLD between wave ABC and wave DEF is the distance GEH, where

$$GE = EH = d \sin \theta$$

Thus the condition for constructive interference from atomic planes can be written as

$$n\lambda = 2d \sin \theta \qquad n = 1, 2, 3, \ldots \tag{10.14}$$

and is known as *Bragg's law*.

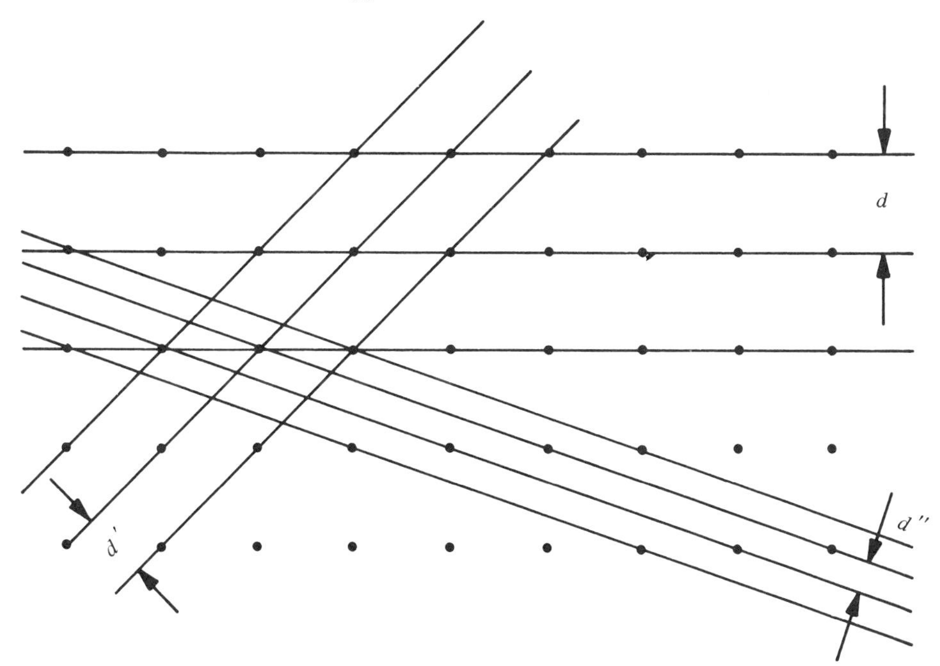

FIGURE 10.10
Three sets of parallel Bragg planes for a lattice structure with $d > d' > d''$

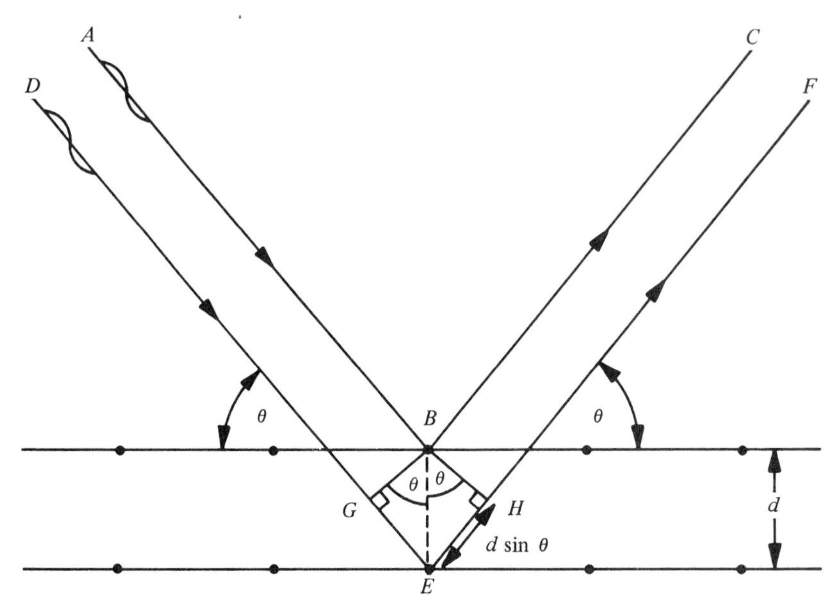

FIGURE 10.11
Bragg scattering from two successive planes

Note that if $2d < \lambda$, no diffraction is possible because $\sin\theta$ would have to be greater than one. For $2d > \lambda$ we can observe different orders of diffraction ($n = 1, 2, \ldots$) at different angles. Equation (10.14) is the basis of all coherent x-ray and electron diffraction effects in crystals. It provides the means of measuring very small wavelengths, comparable to interatomic distances, for if n, d, and θ are known, λ can be computed. On the other hand, if we know λ, n, and θ, we can find the distances d between parallel Bragg planes, and we thereby learn about the structure of crystals. We should also remember that x rays are appropriate radiation for examining crystalline structure because their wavelengths are on the same scale as the interatomic distances being studied. Again the scaling of the ratio d/λ is important. An x-ray diffraction pattern is shown in Figure 10.12.

A few years later, in 1924, Louis de Broglie conjectured that since light was known to have both wave and particle properties, a particle such as an electron might also have wave properties! Several diffraction experiments were soon performed by Davisson and Germer, experiments that verified de Broglie's idea that every moving particle has a wavelength associated with it. The relationship between the speed v of the moving particle and its wavelength is given by the de Broglie

FIGURE 10.12
X-ray diffraction pattern of polycrystalline aluminum. The center is dark because a hole was cut in the photographic plate to allow the strong central beam to pass through it. (Reprinted from R. T. Weidner and R. L. Sells, *Elementary Modern Physics*. Copyright © 1973. Reprinted by permission of Allyn and Bacon, Inc.)

relation

$$\lambda = \frac{h}{mv} \tag{10.15}$$

where h is Planck's constant* and m is the mass of the particle. This finding opened up the way to the use of electrons and neutrons in ongoing research into the structure of solids. From Eq. (10.15) we see that the faster a particle moves, the shorter its wavelength becomes. An electron moving at a speed of 7×10^6 m/sec has a wavelength of 1 Å, which is on the same scale as the atomic spacing in crystals. Thus, electrons at these speeds produce interference patterns when they are used to bombard crystals. Such electron diffraction is used in the electron microscope.

Neutrons can also be used to study crystalline structure. Typically, neutrons used in diffraction experiments are produced in nuclear reac-

*Planck's constant has the value $h = 6.6256 \times 10^{-34}$ joule-second (J-sec).

tors. These neutrons have very high energies and high speeds, and therefore very small wavelengths (less than 1 Å). Consequently, they must be slowed down to have a wavelength nearer 1 Å, the atomic spacing. This is accomplished by passing the neutrons through some material in which they lose kinetic energy during collisions with individual atoms. When the neutrons have been slowed to the point where they have a wavelength of approximately 1 Å, they can be used to study crystalline structure. Conversely, diffraction of neutron (and electron) beams by crystals can provide information on the wavelength (and therefore the speed) of these high-energy particles if a crystal with known lattice structure is used as the diffracting object.

SOUND WAVE DIFFRACTION

Yet another type of wave whose diffraction has significant effects is the sound wave. The wavelengths of audible sound range from 2 cm to 20 m. These wavelengths are much larger than those of visible light used with diffraction gratings, and very much (by a factor of 10^8) larger than the 1-Å wavelengths used in x-ray and atomic particle diffraction. In fact, the dimensions of the wavelengths of sound are comparable to those of typical objects in rooms, of windows, and of door openings. This means that we hear a substantial amount of sound by diffraction. Here again we are primarily interested not in the dimension of λ alone, nor in that of the diffracting object or aperture alone, but in the ratio of wavelength to object size.

As we might expect, there are many similarities between the diffraction of sound waves and the diffraction of light waves. The main difference is one of scale. Figure 10.13a shows the diffraction of a sound wave through an aperture that is shorter than the wavelength of the incoming wave, while Figure 10.13b illustrates diffraction through an opening of roughly the same length as λ. Again, the amount of spreading of the wave depends on the ratio a/λ. Diffraction also takes place whenever a wave simply passes over an edge, and we see in Figure 10.13c the diffraction of a sound wave around a partition. For an object or aperture of a given size, the bending into the "shadow zone" behind the object or aperture will increase with wavelength. A simple experiment will demonstrate this phenomenon. Listen to music coming through an open doorway some distance away. Approach the doorway from one side, not head-on. When still in the shadow zone, you will hear primarily the low-pitched (long wavelengths) sounds. As you come closer to the doorway, the sound becomes more lively, since the higher frequencies (shorter wavelengths) are now detected with greater

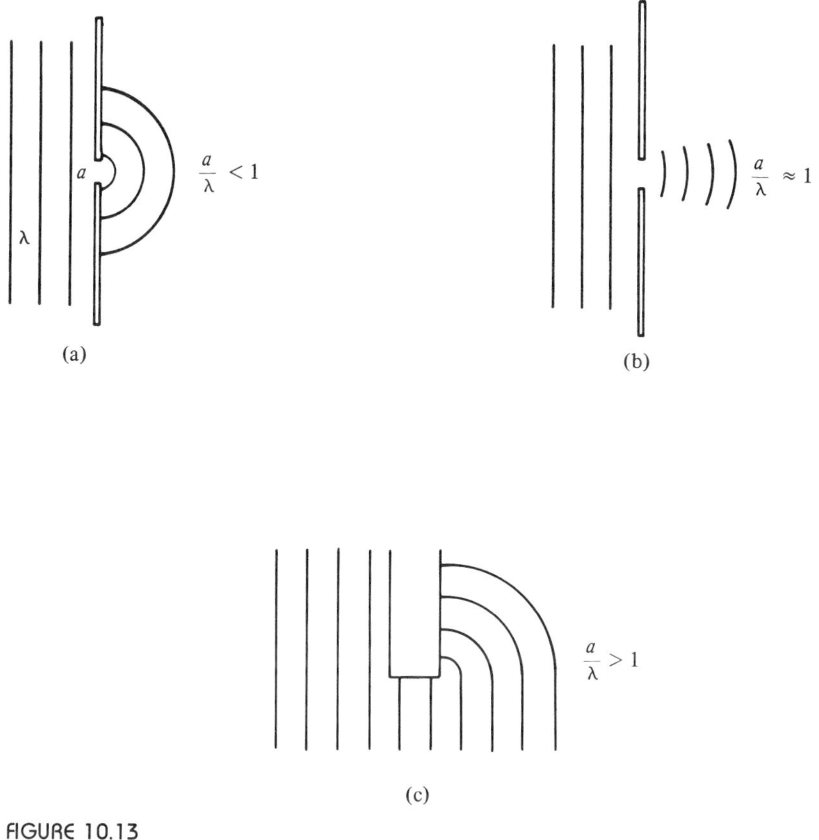

FIGURE 10.13
Diffraction of sound waves

intensity. Designers of auditoriums use this diffraction phenomenon to bring the whole frequency range to as much of the audience as possible. They accomplish this objective by adding to the walls and ceilings a variety of geometric protrusions of varying sizes and shapes, all of which are diffracting devices.

There are many situations in which the diffraction of sound waves is important in the propagation of sound. We will look at two representative cases, the effect of diffraction on the frequency response of a microphone, and the effect of diffraction on the propagation of sound over barriers.

Whether or not the presence of a microphone disturbs the sound field depends on the ratio of the radius r of the microphone to the wavelength of the sound. As discussed previously, the amount of bending of a wave when it strikes an object depends on what the wave "sees." In this case, the larger the radius of the microphone with respect to the wavelength of the incoming sound wave, the more the presence of the microphone disturbs the sound field. The disturbance in the sound field occurs because of greater diffraction of the wave around the microphone. For a cylindrical microphone with the diaphragm (vibrating membrane) at one end, this disturbance of the wave reaches a maximum when the diameter of the microphone is approximately the same as the wavelength, and when the wave propagation is perpendicular to the plane of the diaphragm. Now suppose a wave approaches a microphone at some angle θ as shown in Figure 10.14. The angle θ is called the *angle of incidence*. The wave "sees" a differently shaped diaphragm for different angles of incidence. Therefore, the diffraction of the wave varies with different angles of incidence, and in turn the response of the microphone varies. Again, we note that the change in microphone response with change in θ becomes negligible as the wavelength becomes much larger than the diameter of the microphone. This happens because what the wave "sees" is so small that any change in the "observed" shape of the diaphragm is too small to "see."

Consider now the situation where a barrier is erected between a sound source and a receiver. We see such barriers along heavily traveled highways that are situated adjacent to houses or apartments. Their use provides appreciable noise reduction for those living nearby because sound can reach the receivers only by diffraction over the top and around the ends of the barrier. Sound intensity drops off (that is, is attenuated) with increasing distance between source and receiver, without anything between them. The excess (additional) sound attenuation EA provided by a rigid straight barrier above such normal attenuation of direct line-of-sight propagation can be expressed as

$$EA \sim 20 \log(\sqrt{2\pi N}/\tanh \sqrt{2\pi N}) \tag{10.16}$$

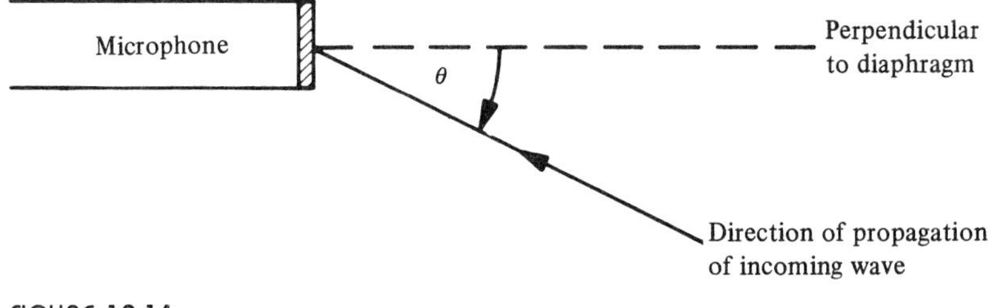

FIGURE 10.14
Definition of the angle of incidence, θ

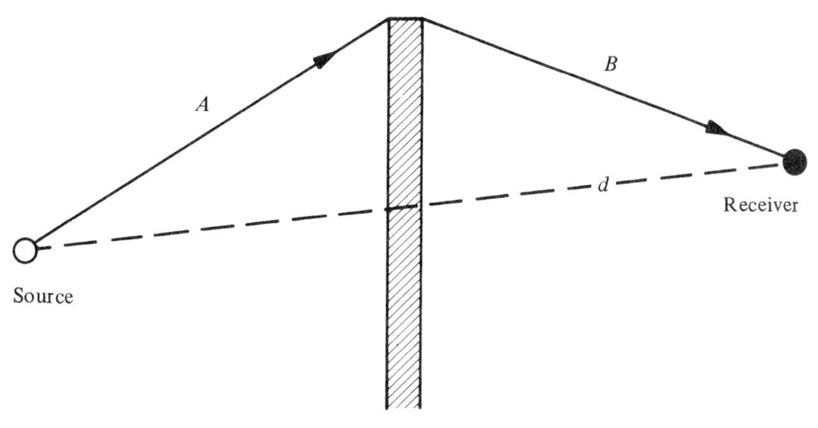

FIGURE 10.15
Geometry showing sound propagation path over a barrier

where N is a dimensionless number, called the Fresnel number, whose meaning we now describe. Since the excess attenuation depends on diffraction of the sound wave, and since we have seen several examples of the dependence of diffraction on the relation between wavelength and dimensions of the diffracting object, we can expect the Fresnel number to depend on these features. Figure 10.15 illustrates the barrier geometry where path $A + B$ is the actual diffracted path of the sound wave as it propagates from source to receiver. The distance d is the direct path between source and receiver without the barrier in place.

The difference in path length between the diffracted path $A + B$ and the direct path d can be written as the path length difference δ

$$\delta = A + B - d \tag{10.17}$$

Then the Fresnel number becomes

$$N = 2\delta/\lambda \tag{10.18}$$

Once again we have a ratio of some length divided by the wavelength controlling the diffraction of a wave. The greater the path length difference, that is, the higher the barrier, the more sound attenuation provided (see Figure 10.15).

SUMMARY

In this chapter we have discussed the relationship between diffraction patterns and the dimensions of wavelengths and diffracting objects.

Different types of diffracting waves were examined, and the similarities between them were discussed. The diffraction patterns of visible light have long been investigated and used in research. Only recently (since 1900), however, has the diffraction of x rays, electrons, and neutrons been understood, and therefore used, in studying crystalline structures.

A brief look at the effects of the diffraction of sound provided an opportunity to examine diffraction involving much larger objects and much longer waves than in the previous cases. We have demonstrated in all these phenomena the close relationship between the dimensions of the interacting parties and the resulting diffraction. Finally, the usefulness of diffraction as a research tool has been stressed throughout the chapter.

Problems

10.1 Three waves converge on the same spot at the same time. They are all of the same frequency, and all are traveling in the x direction. If the amplitudes are 1, 2, and 0.5 mm, respectively, and if the first and second waves are in phase with each other while the third wave is completely out of phase with the first two, find the resultant wave. Sketch all four waves.

10.2 (a) Why in daily experience is the diffraction of sound waves more evident than that of light waves?

(b) For Fraunhofer diffraction through a single slit, what is the effect of increasing (1) the wavelength and (2) the slit width?

10.3 Find the position (θ) of the first-order maximum for yellow light of wavelength 5890 Å when a diffraction grating that has 10^4 rulings uniformly spaced over 2 cm is used.

10.4 (a) Find the position of the first-order maximum for the same yellow light as in Problem 10.3 with a modified diffraction grating. The modified grating has the same number of rulings (10^4) but is half as wide as that in Problem 10.3.

(b) How does the change in θ between the value found in part (a) and that found in Problem 10.3 relate to the ratio d/λ?

10.5 Consider the following situation. Someone is walking down the center of a 6-ft-wide hallway. Ahead are two open doors, one on each side of the hallway directly opposite each other. Sound is emanating from each doorway with a 4000-Hz tone coming from the left and a 5000-Hz tone coming from the right. If these sounds have the same amplitude and phase and emanate from doorways 1 m in width, how far along the hall, beyond the center of the doorways, will the person walk before experiencing the first minimum?

10.6 (a) Partially close the doorway on the right in Problem 10.5 until the opening has been reduced to 0.5 m. Where along the hall does the first minimum occur under these conditions if all other values remain the same as in Problem 10.5?

(b) Discuss the effects of change in doorway width. Would these effects be different if the tone coming from the right were of lower frequency than that coming from the left? Explain.

10.7 (a) First-order Bragg diffraction is observed at an angle of 0.04° with respect to an incident beam of monochromatic x rays. The x rays impinge upon a single KCl crystal and have an energy of 8×10^{-13} J. (The energy is given by $E = hc/\lambda$.) What is the crystal lattice spacing?

(b) How does the lattice spacing compare with the wavelength of the x rays?

10.8 A laboratory experiment is performed using a large-scale model of a crystalline lattice. The lattice consists of small spheres at the corners of densely packed cubes. The cubes have edges 10 cm long. If the "crystal" is irradiated with microwaves whose direction is perpendicular to a cubic face, strong Bragg diffraction is produced by 45° planes when $\theta = 12.2°$. What is the wavelength of the microwaves?

10.9 (a) Equation (10.10) gives the intensity of fringes for two or more slits. Write the analogous equation for a single slit.

(b) If a single slit is illuminated with coherent light, show that $I_\theta = I_{\max}$ when $\theta = 0°$.

10.10 For a thin sound barrier 5 m high, what will the excess attenuation (EA) of sound be for an observer located behind the barrier when the heights of the point source of sound and that of the observer are the same? The horizontal distance between source and observer is 35 m, and the source is located 10 m from the barrier. The source is emitting a 500-Hz tone. Use EA = 20 $\log(\sqrt{2\pi N}/\tanh\sqrt{2\pi N}) + 5$ decibels (dB). (A decibel is a unit of sound pressure level.)

10.11 (a) It is desired to be able to read off values of EA from a graph for values of N between 1 and 100. Plot a curve that will serve this purpose.

(b) Experimental data show that a practical upper limit of about 24 dB of excess attenuation results when a barrier is inserted between a sound source and an observer. This implies a maximum value for N as well. What is that maximum value?

APPENDIX I: ELEMENTARY TRANSCENDENTAL FUNCTIONS

The so-called *elementary transcendental functions* are the trigonometric functions ($\sin x$, $\cos x$), exponential functions (e^x, a^x), logarithmic functions ($\ln x = \log_e x$, $\log_a x$), and hyperbolic functions ($\sinh x$, $\cosh x$). We shall present here some basic results and relationships involving these functions; we are not really intent on presenting derivations and proofs. Some of our results make use of the notation $i = \sqrt{-1}$, this being a central concept in relating, for example, the trigonometric functions to the exponential. (See also p. 133). We note that most of the results presented here can be obtained (by formal manipulation or by arithmetic computations) from the Taylor series for the exponential function

$$e^x = 1 + x + \frac{1}{2!}x^2 + \frac{1}{3!}x^3 + \cdots + \frac{1}{n!}x^n + \cdots \qquad \text{(A1.1)}$$

The exception is the definition of the natural logarithm, which is that $x = \ln y$ if $y = e^x$. The series representation of the logarithm is (for $x \neq -1$ and $|x| \leq 1$)

$$\ln(1 + x) = x - \frac{1}{2}x^2 + \frac{1}{3}x^3 - \cdots \qquad \text{(A1.2)}$$

Further, the natural logarithm can be related to the logarithm "to base 10" by the equation

$$\ln x \cong 2.303 \log_{10} x \qquad \text{(A1.3)}$$

Some properties of the exponential function that are easily demonstrated are that

$$e^{x+y} = e^x e^y \tag{A1.4}$$

and

$$e^{ax} = (e^x)^a \tag{A1.5}$$

Also, the exponential function can be related to the trigonometric functions by calculating

$$\begin{aligned} e^{ix} &= \left(1 - \frac{1}{2!}x^2 + \frac{1}{4!}x^4 - \cdots\right) + i\left(x - \frac{1}{3!}x^3 + \frac{1}{5!}x^5 - \cdots\right) \\ &= \cos x + i \sin x \end{aligned} \tag{A1.6}$$

Note that in computing the result in Eq. (A1.6) we have used the series definitions of the trigonometric functions [Eqs. (4.11)].

The hyperbolic functions are defined in terms of the exponential function:

$$\sinh x = \frac{1}{2}(e^x - e^{-x}) \tag{A1.7}$$

$$\cosh x = \frac{1}{2}(e^x + e^{-x}) \tag{A1.8}$$

However, we can also relate these hyperbolic functions directly to the trigonometric functions, either by using Eq. (A1.6) in the foregoing definitions, or by calculating $\sin ix$ and $\cos ix$ with Eqs. (4.11). In either case we would find that

$$\begin{aligned} \sinh ix &= i \sin x \qquad & \sin x &= -i \sinh ix \\ \sin ix &= i \sinh x \qquad & \sinh x &= -i \sin ix \end{aligned} \tag{A1.9}$$

and

$$\begin{aligned} \cosh ix &= \cos x \qquad & \cos x &= \cosh ix \\ \cos ix &= \cosh x \qquad & \cosh x &= \cos ix \end{aligned} \tag{A1.10}$$

Although the structures of the trigonometric and hyperbolic functions appear to be very similar, the behavior of these two classes of functions is very different. (Oh, what a difference an i makes!) The trigonometric functions are periodic, of period 2π, and their absolute values are always bounded by unity, that is, $-1 \leq \sin x, \cos x \leq 1$. The hyperbolic functions are not periodic. The hyperbolic sine increases monotonically for all values of its argument. The hyperbolic cosine does so for positive values of its argument and is symmetric about the origin. Both hyperbolic functions become infinitely large as x becomes infinite. Table A1.1 presents some basic features of all of the elementary functions.

Some derivatives and integrals of the elementary functions are

$$\frac{d}{dx}\sin x = \cos x \qquad \frac{d}{dx}\cos x = -\sin x$$
$$\frac{d^2}{dx^2}\sin x = -\sin x \qquad \frac{d^2}{dx^2}\cos x = -\cos x \tag{A1.11}$$

$$\frac{d}{dx}e^x = e^x \qquad \frac{d^n}{dx^n}e^x = e^x \tag{A1.12}$$

$$\frac{d}{dx}\sinh x = \cosh x \qquad \frac{d}{dx}\cosh x = \sinh x$$
$$\frac{d^2}{dx^2}\sinh x = \sinh x \qquad \frac{d^2}{dx^2}\cosh x = \cosh x \tag{A1.13}$$

$$\frac{d}{dx}\ln x = \frac{1}{x} \qquad \int \ln x = x\ln x - x \tag{A1.14}$$

Table A1.1
Features of the Elementary Transcendental Functions

	Value at $x = 0$	Value as $x \to \infty$	Behavior
$\sin x$	0	$\|\sin x\| \leq 1$	Periodic
$\cos x$	1	$\|\cos x\| \leq 1$	Periodic
e^x	1	$\to \infty$	Monotonic
$\sinh x$	0	$\to \infty$	Monotonic (all x)
$\cosh x$	1	$\to \infty$	Monotonic ($x > 0$)
$\ln x$	$-\infty$	$\to \infty$	Monotonic
$\log_{10} x$	$-\infty$	$\to \infty$	Monotonic

APPENDIX II: THE DIFFERENTIAL EQUATION $dN/dt = \lambda N$

The differential equation $dN/dt = \lambda N$ is central to the discussion of exponential growth. It is a *first-order* differential equation, that is, the highest-order derivative that appears is of first order. As a consequence, there will be in the general solution one arbitrary constant, this constant arising from the fact that this first-order differential equation needs to be integrated once. Another way of looking at this situation is that the equation gives the rate of growth of the population dN/dt in terms of the population N itself. A complete solution ought to be able to accommodate any initial value of the population, hence there ought to be one arbitrary constant.

We also note that the differential equation of population growth [Eq. (8.1)]

$$\frac{dN(t)}{dt} = \lambda N(t) \tag{A2.1}$$

has *constant coefficients*, that is, the multipliers of $N(t)$ and its derivative are constant (λ and 1, respectively). Putting this notion together with one of the properties of exponential functions, namely that [see Eqs. (A1.12)]

$$\frac{d^n}{dt^n} e^{\alpha t} = \alpha^n e^{\alpha t} \tag{A2.2}$$

leads us to suggest that we seek a solution to Eq. (A2.1) in the form

$$N(t) = Ae^{\alpha t} \tag{A2.3}$$

where α and A are constants that remain to be determined. If we substitute Eq. (A2.3) into Eq. (A2.1) we find that

$$A\alpha e^{\alpha t} = \lambda A e^{\alpha t}$$

or

$$(\alpha - \lambda)Ae^{\alpha t} = 0 \qquad \text{(A2.4)}$$

Clearly we do not want $A = 0$, nor do we expect $e^{\alpha t}$ to equal zero for all time. Hence, $\alpha = \lambda$, and our solution is

$$N(t) = Ae^{\lambda t} \qquad \text{(A2.5)}$$

where the constant A will be determined by the *initial conditions*. Thus, if the population at time $t = 0$ is $N(0) = N_0$, it follows that $A = N_0$, and

$$N(t) = N_0 e^{\lambda t} \qquad \text{(A2.6)}$$

In principle we do not have to determine this constant at $t = 0$. For example, we might specify that $N(t_0) = N_0$, from which we find

$$N_0 = Ae^{\lambda t_0}$$

or

$$A = N_0 e^{-\lambda t_0} \qquad \text{(A2.7)}$$

and

$$N(t) = N_0 e^{-\lambda t_0} e^{\lambda t} = N_0 e^{\lambda(t - t_0)} \qquad \text{(A2.8)}$$

This obviously defines a population that is increasing through N_0 at $t = t_0$, but is less than N_0 for $t < t_0$.

Finally, we note that all of the preceding manipulations are valid for $\lambda < 0$ as well as for $\lambda > 0$. The interpretation would differ, since we would be talking about exponential decay ($\lambda < 0$) rather than growth ($\lambda > 0$), but the mathematics would not change.

APPENDIX III: THE DIFFERENTIAL EQUATION $m\, d^2x/dt^2 + kx = F(t)$

The differential equation

$$m \frac{d^2x}{dt} + kx = F(t) \tag{A3.1}$$

is the basic equation governing the response of simple harmonic oscillators (Chapter 6), a special case of which is the linear model of the pendulum [see Eq. (5.18), for example]. This equation also has *constant coefficients* (see Appendix II), and is of *second order*. Equation (A3.1) is also *inhomogeneous* because of the forcing function $F(t)$ on the right-hand side. To simplify the discussion, let us first consider the *homogeneous* differential equation

$$m \frac{d^2x}{dt^2} + kx = 0 \tag{A3.2}$$

Because of the constant coefficients k and m, we are led to try a solution of the form of Eq. (A2.3):

$$x(t) = Ae^{\alpha t} \tag{A3.3}$$

Substitution quickly leads to

$$(m\alpha^2 + k)Ae^{\alpha t} = 0 \tag{A3.4}$$

from which it follows that $\alpha^2 = -k/m$, or

$$\alpha = \pm\sqrt{-1}\,\sqrt{\frac{k}{m}} = \pm i\omega_0 \tag{A3.5}$$

In Eq. (A3.5) we have noted that $i = \sqrt{-1}$, and we have defined a constant [see Eq. (6.7)]

$$\omega_0 = \sqrt{\frac{k}{m}} \tag{A3.6}$$

We read from Eq. (A3.5) that there are two solutions to the homogeneous oscillator equation, each with its own arbitrary constant,

$$x_1 = A_1 e^{i\omega_0 t} \qquad x_2 = A_2 e^{-i\omega_0 t} \tag{A3.7}$$

The general solution to Eq. (A3.2) is the *sum* of these two solutions,

$$x(t) = A_1 e^{i\omega_0 t} + A_2 e^{-i\omega_0 t} \tag{A3.8}$$

We can put Eq. (A3.8) into a more familiar form by rewriting it with the aid of Eq. (A1.6):

$$\begin{aligned} x(t) &= A_1 (\cos \omega_0 t + i \sin \omega_0 t) + A_2 (\cos \omega_0 t - i \sin \omega_0 t) \\ &= (A_1 + A_2) \cos \omega_0 t + i(A_1 - A_2) \sin \omega_0 t \\ &= B_1 \cos \omega_0 t + B_2 \sin \omega_0 t \end{aligned} \tag{A3.9}$$

where in the last result we have replaced the arbitrary constants A_1 and A_2 by another (equivalent) set of arbitrary constants. It is easily verified by direct substitution that Eq. (A3.9) satisfies Eq. (A3.2).

The solution represented by Eq. (A3.9) is often called the *transient solution* to the differential equation because it can be said to straightforwardly represent the initial conditions. Thus, if $x(0) = x_0$, and $dx(0)/dt = \dot{x}_0$, it is easy to show that $B_1 = x_0$ and $B_2 = \dot{x}_0/\omega_0$, so

$$x(t) = x_0 \cos \omega_0 t + \frac{1}{\omega_0} \dot{x}_0 \sin \omega_0 t \tag{A3.10}$$

In a (physically) real oscillator there would be some damping, which over a period of time would damp out the effect of the initial conditions—and so the origin of the label "transient solution." This damping out of the initial conditions is even more evident when we look at a particular inhomogeneous equation (and its solution).

Let us consider the inhomogeneous oscillator [Eq. (A3.1)], where the forcing function is a periodic oscillation defined by

$$F(t) = F_0 \cos \omega t, \qquad \omega \neq \omega_0 \tag{A3.11}$$

where F_0 and ω are arbitrary. The reason for the restriction $\omega \neq \omega_0$ will become clear quite soon. Also, we note that an implicit condition of what we are about to do is that the forcing function is applied for all time, and we are seeking a *particular* solution that describes the response of the oscillator to the periodic force $F(t)$ for all time. This particular solution is thus called the *steady-state* response, and it is entirely independent of the initial conditions (and thus of the transient solution).

As a trial steady-state solution let us assume that

$$x(t) = C \cos \omega t \tag{A3.12}$$

where C is an (as yet) undetermined constant. By direct substitution into

$$m \frac{d^2x}{dt^2} + kx = F_0 \cos \omega t, \qquad \omega \neq \omega_0 \tag{A3.13}$$

we get

$$-m\omega^2 C \cos \omega t + kC \cos \omega t = F_0 \cos \omega t$$

or

$$(k - m\omega^2) C \cos \omega t = F_0 \cos \omega t$$

or

$$C = \frac{F_0}{k - m\omega^2} = \frac{F_0}{m(\omega_0^2 - \omega^2)} = \frac{F_0/k}{1 - \omega^2/\omega_0^2} \tag{A3.14}$$

and

$$x(t) = \frac{F_0/k}{1 - \omega^2/\omega_0^2} \cos \omega t \tag{A3.15}$$

We see very clearly from Eqs. (A3.14) and (A3.15) why we need the restriction $\omega \neq \omega_0$, for if the forcing frequency ω has a value almost equal to that of the natural frequency ω_0, we appear to get an indefinitely large response, that is, $x(t) \to \infty$. This is the condition of resonance. In fact, the response of a real oscillator does become large, but not infinitely so, because damping prevents an infinite response. A solution for the case $\omega = \omega_0$, with and without damping, is beyond our scope here, although some results for these cases are given in Chapter 6 in the section before the summary.

BIBLIOGRAPHY AND REFERENCES

The alphabetical listing of works that have been consulted is followed by an ordering that is keyed to specific chapters.

1. T. Au and T. E. Stelson, *Introduction to Systems Engineering*. Addison-Wesley, Reading, Mass., 1969.
2. A. A. Bartlett, "The Exponential Function, Parts I–VIII," *The Physics Teacher*: I, October 1976 (p. 393); II, November 1976 (p. 485); III, January 1977 (p. 37); IV, February 1977 (p. 98); V, April 1977 (p. 225); VI, January 1978 (p. 23); VII, February 1978 (p. 92); VIII, March 1978 (p. 158).
3. R. A. Becker, *Introduction to Theoretical Mechanics*. McGraw-Hill, New York, 1954.
4. L. Beranek, *Noise and Vibration Control*. McGraw-Hill, New York, 1971.
5. M. Braun, *Differential Equations and Their Applications*. Springer-Verlag, New York, 1975.
6. J. C. Burkill, *A First Course in Mathematical Analysis*. Cambridge Univ. Press, London and New York, 1967.
7. W. Burns, *Noise and Man*. Lippincott, Philadelphia, 1973.
8. R. H. Cannon, Jr., *Dynamics of Physical Systems*. McGraw-Hill, New York, 1967.
9. R. E. Chandler, R. Herman, and E. W. Montroll, "Traffic Dynamics: Studies in Car Following," *Operations Research* **6**(2), 165–184 (1958).
10. A-M. Chung, *Linear Programming*. Merrill, New York, 1963.
11. J. Cohen, *Diffraction Methods in Materials Science*. Macmillan, New York, 1966.
12. K. C. Crandall and R. W. Seabloom, *Engineering Fundamentals in Measurements, Probability, Statistics, and Dimensions*. McGraw-Hill, New York, 1970.
13. J. H. Engel, "A Verification of Lanchester's Law," *Operations Research* **2**(2), 163–171 (1954).
14. A. A. Ezra, "Scaling Laws and Similitude Requirements for Valid Scale Model Work," in *Use of Models and Scaling in Shock and Vibration* (W. E. Baker, ed.). American Society of Mechanical Engineers, New York, 1963.

15. R. P. Feynman, R. B. Leighton, and M. Sands, *The Feynman Lectures on Physics,* Vols. I and II. Addison-Wesley, Reading, Mass., 1963.
16. J. W. Forrester, *World Dynamics.* Wright-Allen, Cambridge, Mass., 1971.
17. M. Francon, *Diffraction, Coherence in Optics.* Pergamon, Oxford, 1966.
18. R. J. Giglio and R. Wrightington, "Methods for Apportioning Costs among Participants in Regional Systems," *Water Resources Research* **8**(5), 1133–1144 (1972).
19. B. R. Gossick, *Hamilton's Principle and Physical Systems.* Academic Press, New York, 1967.
20. S. J. Gould, "Size and Shape," *Harvard Magazine* October, pp. 43–50 (1975).
21. R. Haberman, *Mathematical Models*, Prentice-Hall, Englewood Cliffs, N.J., 1977.
22. D. Halliday and R. Resnick, *Fundamentals of Physics.* Wiley, New York, 1970.
23. S. I. Hayakawa, *Language in Thought and Action.* Harcourt, New York, 1949.
24. R. Herman, E. W. Montroll, R. B. Potts, and R. W. Rothery, "Traffic Dynamics: Analysis of Stability in Car Following," *Operations Research* 7(1), 86–106 (1959).
25. J. P. Holman, *Experimental Methods for Engineers.* McGraw-Hill, New York, 1971.
26. R. W. Hornbeck, *Numerical Methods.* Quantum, New York, 1975.
27. G. W. Housner and D. E. Hudson, *Applied Mechanics: Dynamics.* Van Nostrand-Reinhold, New York, 1959.
28. D. Huff, *How to Lie with Statistics.* Norton, New York, 1954.
29. F. A. Jenkins and H. E. White, *Fundamentals of Optics.* McGraw-Hill, New York, 1950.
30. L. B. W. Jolley, *Summation of Series.* Dover, New York, 1961.
31. T. von Karman, *Aerodynamics.* McGraw-Hill, New York, 1963.
32. J. G. Kemeny, *A Philosopher Looks at Science.* Van Nostrand-Reinhold, New York, 1959.
33. L. E. Kinsler and A. R. Frey, *Fundamentals of Acoustics.* Wiley, New York, 1962.
34. S. J. Kline, *Similitude and Approximation Theory.* McGraw-Hill, New York, 1965.
35. F. W. Lanchester, *Aircraft in Warfare; the Dawn of the Fourth Arm.* Constable, London, 1916. See also J. R. Newman (ed.), *The World of Mathematics.* Simon and Schuster, New York, 1956.
36. H. L. Langhaar, *Dimensional Analysis and Theory of Models.* Wiley, New York, 1951.
37. R. E. Lapp and H. L. Andrews, *Nuclear Radiation Physics.* Prentice-Hall, Englewood Cliffs, N.J., 1954.
38. A. M. Lee, *Applied Queueing Theory.* St. Martin's Press, New York, 1966.
39. J. C. Liebman, J. W. Male, and M. Wathne, "Minimum Cost in Residential Refuse Vehicle Routes," *Journal of the Environmental Engineering Division* (Proceedings of the ASCE, Vol. 101, No. EE3, June 1975), pp. 399–412.
40. C. Lipson and N. J. Sheth, *Statistical Design and Analysis of Engineering Experiments.* McGraw-Hill, New York, 1973.

41. E. Magrab, *Environmental Noise Control*. Wiley, New York, 1975.
42. D. P. Maki and M. Thompson, *Mathematical Models and Applications*. Prentice-Hall, Englewood Cliffs, N.J., 1973.
43. D. H. Meadows, D. L. Meadows, J. Randers, and W. W. Behrens III, *The Limits to Growth*. Universe Books, New York, 1972.
44. H. J. Miser, "Introducing Operational Research," *Operational Research Quarterly* **27** (3), 655–670 (1976).
45. G. Murphy, *Similitude in Engineering*. Ronald Press, New York, 1950.
46. E. C. Pielou, *An Introduction to Mathematical Ecology*. Wiley (Interscience), New York, 1969.
47. J. A. Roberson and C. T. Crowe, *Engineering Fluid Mechanics*. Houghton-Mifflin, Boston, 1975.
48. M. F. Rubinstein, *Patterns of Problem Solving*. Prentice-Hall, Englewood Cliffs, N.J., 1975.
49. H. Schenck, Jr., *Theories of Engineering Experimentation*. McGraw-Hill, New York, 1968.
50. L. Schwartz and J. Cohen, *Diffraction from Materials*. Academic Press, New York, 1977.
51. F. W. Sears and M. W. Zemansky, *Modern University Physics*. Addison-Wesley, Reading, Mass., 1960.
52. J. Singh, *Great Ideas of Operations Research*. Dover, New York, 1968.
53. J. M. Smith, *Mathematical Ideas in Biology*, Cambridge Univ. Press, London and New York, 1968.
54. R. M. Stark and R. L. Nicholls, *Mathematical Foundations for Design*. McGraw-Hill, New York, 1972.
55. G. W. Swenson, Jr., *Principles of Modern Acoustics*. Boston Technical Publishers, Cambridge, Mass., 1965.
56. E. S. Taylor, *Dimensional Analysis for Engineers*. Oxford (Clarendon Press), London and New York, 1974.
57. H. Theil, J. C. G. Boot, and T. Kloek, *Operations Research and Quantitative Economics*. McGraw-Hill, New York, 1965.
58. D. W. Thompson, *On Growth and Form*. Cambridge Univ. Press, London and New York, 1969. (Abridged edition, J. T. Bonner, ed.)
59. S. P. Timoshenko and B. F. Langer, "Stresses in Railroad Track," *Transactions of the American Society of Mechanical Engineers* **54**, 277–293 (1932).
60. P. A. Tipler, *Foundations of Modern Physics*. Worth, New York, 1969.
61. C. Toregas and C. ReVelle, "Binary Logic Solutions to a Class of Location Problem," *Geographical Analysis* **5** (7), 145–155 (1973).
62. H. M. Wagner, *Principles of Operations Research*. Prentice-Hall, Englewood Cliffs, N.J., 1969.
63. J. A. Wattleworth, "Traffic Flow Theory," in *Transportation and Traffic Engineering Handbook*, (J. E. Baerwald, ed.). Prentice-Hall, Englewood Cliffs, N.J., 1976.
64. M. R. Wehr and J. A. Richards, Jr., *Physics of the Atom*. Addison-Wesley, Reading, Mass., 1960.
65. R. T. Weidner and R. L. Sells, *Elementary Modern Physics*. Allyn and Bacon, Boston, 1973.
66. F. A. White, *Our Acoustic Environment*. Wiley, New York, 1975.
67. R. M. Whitmer, *Electromagnetics*. Prentice-Hall, Englewood Cliffs, N.J., 1962.

68. M. Wohl and B. V. Martin, *Traffic System Analysis*. McGraw-Hill, New York, 1967.
69. H. D. Young, *Statistical Treatment of Experimental Data*. McGraw-Hill, New York, 1962.

Specific chapter references are listed below.

Chapter	References
1	23, 32, 44, 48
2	36, 47, 49, 56
3	7, 14, 20, 31, 33, 34, 45, 53, 58
4	3, 6, 12, 21, 25, 26, 28, 30, 40, 49, 51, 69
5	15, 27
6	8, 15, 19, 55, 64, 67
7	9, 21, 24, 63, 68
8	2, 5, 13, 16, 35, 37, 43, 46, 53
9	1, 10, 18, 38, 39, 42, 52, 54, 57, 61, 62
10	4, 11, 15, 17, 22, 29, 33, 41, 50, 59, 60, 65, 66

INDEX

A

Absolute error, 64n
Acceleration
  angular, of wheel, 42–44
  of particle in magnetic field, 126
  of pendulum, 90–92, 94
Accuracy, 65
Acoustic resonator, 37–38, 122–125
Acoustics
  frequency analysis, 44–46
  scaling, in human speech and hearing, 36–38
  sound wave diffraction, 234–237
Addition, and significant figures, 61
Adiabatic gas law, 123
Aerodynamics, of birds in flight, 33–36
Algebraic approximation, 59–63
Angle of incidence, 236
Approximation technique, 51–81
  algebraic and significant figures, 59–63
  averaging, 66–69
  binomial expansion, 54–57
  curve fitting, 69–77
  errors, 63–66
  mode validation, 77–78
  Taylor series, 51–53
  trigonometric series, 57–59
Area, surface, of cube, 29
Asymptotic behavior, 162–163, 171–172
Automobile suspension system, 119–122
Average, 66–69

B

Bar chart, *see* Histogram
Basic method, of dimensional analysis, 21–24
  pendulum motion, 88
Beams, theory of, 2–3
Binomial expansion, 54–57
Boundary conditions, and scaling, 39
Bragg plane, 231–232
Bragg's law, 231
Buckingham's Pi theorem, *see* Pi theorem of Buckingham
Building
  scaling, 31–33
  vibration, 116–119

C

Calculus, optimization via, 193–198
Cantilever beam, vibration, 118–119
Capacitance, 128
Capacitor, 127–129, 169–172
Car-following model, 139, 147–156
Catenary, equation of, 57
Cathedral, scaling, 31–33
Centripetal acceleration, 92

Circuit, electrical, *see* Electrical circuit
Circular frequency
  of pendulum, 94
  of spring–mass system, 115
Compound interest, 173–175
Computer, use in plotting graphs, 77
Conceptual world, in scientific method, 4–5
Cone, volume, 30–31
Conservation of cars, in traffic flow theory, 143–144
Conservation of energy
  in pendulums, 97–100
  in spring–mass system, 114
Constant coefficient, 245, 247
Constructive interference, 218, 220, 231
Continuity equation, of fluid mechanics, 144
Continuous curve, 75–76
Continuum hypothesis, 76, 140–142
Contraction of materials, 38
Cost–benefit analysis, 191–192
Coulomb, 125
Coupled system of growth
  Lanchester's law, 184–188
  Lotka–Volterra equations, 179–184
Crystals, and x-ray diffraction, 230–234
Cube, scaling, 28–29
Curve fitting, 69–77
Cyclotron frequency, 127
Cylinder, volume, 29–30

D

Damping
  automobile suspension system, 120
  friction force, 100–101
  oscillator, 134–135, 248–249
  pendulum, 101
  spring–mass system, 113–114
de Broglie relation, 232–233
Decay, exponential, 159, 162
  capacitor discharge, 169–171
  Lanchester's law, 184–188
  Lotka–Volterra equations, 179–184
  radioactive decay, 167–169
Density, in traffic flow models, 141–155
Derived quantity, 12
Destructive interference, 218, 223
Differential equations, 245–249
  for energy as function of time, 102
  of motion, 91–92
Diffraction, 217–239
  geometry of, 218–225
  sound wave, 234–237
  x-ray, 230–234
Diffraction grating, 226–230
Dimension, 12–13
Dimensional analysis, 11–27
  motivation, 14–16
  of pendulum motion, 87–89, 92–95
  process of, 17–24
Dimensional homogeneity, 13–14
Dimensionless group, 17
  spring–mass system, 19–20
  in submarine drag force problem, 17–18
Discrete traffic flow model, 142
Dissipation of energy
  in pendulums, 100–103
  spring–mass system, 114
Dissipative force, *see* Friction force
Division, and significant figures, 61–62
Double slit diffraction, 218, 220–222
Doubling time, 165
Drag force, on submarine, 14–16, 18
Driving frequency, of oscillator, 130–132
Dynamic programming, 203

E

e, 164
Eardrum, scale effects, 36–37
Elastic spring, 112
Electric current, 128
Electrical circuit, 127–130
  capacitor charging, 171–172
  capacitor discharge, 169–171
Electrical–mechanical analogy, 127–130
Electron, 232–234
Elementary particle
  diffraction, 230–234
  in magnetic field, 125–127
Elementary transcendental function, 241–243
Elliptic integral, 105–106
Elliptical curve, 181–182, 184
Energy
  of pendulums, 97–103
  of spring–mass system, 114
English system of units, 25

Error, 63–66
  and choice of scale, 42–44
  and significant figures, 60
Excitation frequency, of oscillator, 133
Expansion, thermal, *see* Thermal expansion
Exponential function, 164–167, 241–243
Exponential model, 159–189
  capacitor discharge, 169–171
  compound interest, 173–175
  highways, growth in demand for, 175–176
  inflation, 172–173
  Lanchester's law, 184–188
  Lotka–Volterra equations, 179–184
  population growth, 177–179
  radioactive decay, 167–169
Extrapolation, 69

F

Falling body in vacuum, 21–22
Far-field approximation, 57
Filter, acoustic, 44–45
First-order differential equation, 245–246
Flight of birds, 33–36
Flow network, 209, 211
Forcing frequency, of oscillator, 130–132
Forcing function, 130, 132, 247–249
Fraunhofer diffraction, 218
Free-body diagram, of pendulum, 89–91
Free fall, in vacuum, 21–22
Frequency
  acoustic resonator, 122, 125
  analysis, 44–46
  of buildings, 119
  of eardrum, 36–37
  forcing or driving frequency, 130–132
  of oscillator, 115–116
  of particle in magnetic field, 127
  of spring–mass system, 20
  of vocal cords, 37–38
Fresnel number, 237
Friction force, 100–101
Functional equation, 21–22
Fundamental diagram of road traffic, 142–147
Fundamental quantity, 12

G

Graph
  and choice of scale, 47–48
  curve fitting, 69–77
Gravitational constant, 14, 23–24
Gravitational force
  on pendulum, 89–90
  of spring–mass system, 116
Growth, exponential, 159, 162
  capacitor charging, 171–172
  compound interest, 173–175
  highways, demand for, 175–176
  inflation, 173
  Lanchester's law, 184–188
  limited growth, 177–179
  Lotka–Volterra equations, 179–184
  population, 177–179, 245–246
Growth curve, 159, 178

H

Half-life, 165, 167–168
Hearing, scale effects, 36–37
Heat loss argument, for power of birds in flight, 35
Helmholtz resonator, *see* Acoustic resonator
Hertz, 36
Highways, growth in demand for, 175–176
Histogram, 72–75
Homogeneous differential equation, 247–248
Hydrodynamics, 140
Hyperbolic function, 39, 185, 241–243

I

Impedance, 132–135
Implicit solution, 105
Inductor, 128–129
Inertia, rotational, of wheel, 42–44
Inflation, 172–173
Inhomogeneous differential equation, 247–248
Integer programming, 203–204
Interest, *see* Compound interest
Interference, of waves, 218, 220–222, 225–228, 230–231
Interpolation, 69

K

Kinetic energy, 97–98, 101, 114
Kirchhoff's voltage law, 129, 169

L

Lanchester's law, 184–188, 190
Least squares, method of, 69–72
Length, measurement of, 12–13
Lift force, of birds in flight, 33–34
Light waves, 217–218, 220–223, 227–230
Limb velocity, and animal size, 35
Limited growth, 177–179
Linear car-following model, 147–154
Linear expansion, 59–60
Linear oscillator, 112–135
  acoustic resonator, 122–125
  automobile suspension system, 119–122
  building vibration, 116–119
  electrical circuit, 127–130
  particle in magnetic field, 125–127
  resonance and impedance, 130–135
Linear pendulum model, 95–98
Linear programming, in operations research, 190–216
Linear proportionality, 29–30
Linearity, 29–30
Load, wing, of birds in flight, 33–34
Location analysis, 213
Logarithmic function, 241–243
Logarithms, 164–165, 241
Logistic growth, 178–180
Lotka–Volterra equations, 179–184

M

Macroscopic traffic flow theory, 139–147
Magnetic induction, 125
Manufacturing problem, 200–203
Mass-controlled frequency, 134
Mathematical model, 1–7
  validation, 77–78
Mean, 67–69
Measurements
  approximation in, 60–62
  errors in, 63–66
  numerical measure, 11–13
Mechanics, scale and limits in, 39–41
Median, 67–68
Metric system of units, 25
Microphone, and sound wave diffraction, 235–236
Microscopic traffic flow model, *see* Car-following model
Military science
  Lanchester's law, 184–188
  operations research, 190
Mistake, 65
Mode, 67–68
Model, mathematical, *see* Mathematical model
Monochromatic diffraction, 218, 228
Motion, equations of, 89–92, 110–111
  and time rate of change of energy, 99
Multiplication, and significant figures, 61–62
Muscle force, and power availability of birds in flight, 35

N

Narrow-band frequency analysis, 45
Natural logarithm, 164–165, 241
Network analysis, 208–213
  transporation problem, 204–208
Neutron, 233–234
Newton, 125
Newtonian mechanics, scale and limits in, 39–41
Newton's law of motion, 89–90, 92
  in spring–mass oscillator, 112
Noise control, 236–237
Nonlinear car-following model, 154–156
Nonlinear pendulum model, 99–100, 103–107
Nonlinear programming, 203
Nonlinear term, 92
Nonnegativity constraint, 201–202

O

Observation, in scientific method, 4
Ohm's law, 128
Operations resarch, 190–216
  Lanchester's law, 184–188
  network analysis, 208–213
  optimization, 193–204
  transportation problem, 204–208

Optimization, 190–191
via calculus, 193–198
via linear programming, 198–204
Oscillation
in car-following model, 152–153
in host–parasite populations, 181
Oscillator, *see also* Linear oscillator; Spring–mass system
differential equation, 248–249
Lotka–Volterra equations, 183–184
simple harmonic, 95, 112–113
Oxygen supply argument, for power of birds in flight, 35

P

Parasite–host equations, *see* Lotka–Volterra equations
Particle, elementary, *see* Elementary particle
Passive electrical circuit elements, 127–128
Pendulum, 85–111
dimensional analysis, 87–89, 92–95
energy considerations, 97–103
equations of motion, 89–92
linear model, 95–97
nonlinear model, 103–107
period of, 13–14, 86–87, 97
Percentage error, 64–65
Periodic motion, 85, 87
Perturbation, 182–183
Phase, of waves, 218
Physical quantity, 11–13
Pi theorem of Buckingham, 17–21
Planar wave, 218
Population growth, 159–163, 177–179
differential equation of, 245–246
semilogarithmic plot, 167
Potential, electric, 54–56
Potential energy, 97–99, 101, 114
Pound force, 25
Power availability, of birds in flight, 35–36
Precision, 65
Predator–prey equations, *see* Lotka–Volterra equations
Prediction, in scientific method, 4–5
Primary quantity, 12
Product-mix problem, 200–203

Q

Quantity, physical, *see* Physical quantity

R

Radiation, 217, *see also* specific type of radiated wave
Radioative decay, 167–169
Railroad car wheel vibration, 223–225
Random error, 64
Rational equation, 13
RC circuit, 169–172
Relative error, 64n
Relativity, general theory of, scale and limits in, 39–41
Resistance, 128
Resistor, 128–129
Resonance, 132–135, 249
Revolution of bodies in space, 23–24
Reynolds number, 16
RLC circuit, 129
Rotational inertia, 42–44
Rounding off of numbers, 62
Routing problem, 213

S

Sag, of stretched string, 57–58
Scaling, 28–50
choosing a scale, 41–47
and diffraction, 217
in exponential growth, 160–162
and interference patterns, 228
pendulum, analysis of, 93–94
size and function, 33–38
size and limits, 38–41
size and shape, 28–33
in traffic flow model, 141
Scheduling problem, 204
Scientific method, 3–5
operations research, 190–191
Second-order differential equation, 247–249
Self-inductance, 129
Semilogarithmic plot, 166–167
Sensitivity of measuring device, 65–66
Shock absorber, of automobile, 120
SI system of units, 25
Significant figures, 60–63
Single slit diffraction, 221–223, 228–230

Size
  and function, 33–38
  and limits, 38–41
  and shape, 28–33
Skyscraper, vibration, 116–119
Sound waves, 234–237
Speech, scale effects, 36–38
Speed, *see* Velocity
Sphere, scaling, 29
Spring–mass system, 19–20, 22, 112–116
  automobile suspension system, 119–122
Standard, 12
Standard deviation, 68–69
Statistics
  averaging, 66–69
  errors in, 63–66
Steady-state response, 131, 249
Stiffness-controlled frequency, 133
Straight line, plotting of, 69–73
Strength of materials, and building vibration, 118–119
Submarine, drag force, 14–16, 18
Subtraction, and significant figures, 61
Superposition, of waves, 218–219
Surface, scaling, 29, 32–33
Systematic error, 63–64

T

Taylor series, 51–53, 241
Thermal expansion, 38–39, 59–60
Time rate of change of energy, 98–99, 101–103
Time scale, for pendulum problem, 94–95
Traffic flow model, 139–158
  car-following, 147–156
  macroscopic, 140–147
Transient solution, 131, 248
Transportation problem, 204–208
  network analysis, 208–213
Trigonometric function, 241–243
Trigonometric series, 57–59
Turbulent flow, 16

U

Unit, of quantity, 12–13, 24–26

V

Velocity
  angular, of pendulum, 97
  of animals, 35
  and drag force on submarine, 14–16
  in traffic flow models, 140–155
Vibration
  building, 116–119
  railroad car wheel problem, 223–225
Viscosity of water, and drag force, 14
Viscous friction force, 14, 101
Vocal cords, scale effects, 37–38
Volume
  of cone, 30–31
  of cylinder, 29–30
  scaling, 29–33

W

Warfare
  Lanchester's law, 184–188
  operations research, 190
Wavelength
  and diffraction patterns, 217
  measurement of, 228, 232
  sound waves, 234
  x rays, 230
Waves, *see* Diffraction, and specific type of wave
Wheel, rotational inertia, 42–44
Wing load, of birds in flight, 33–34
Work, muscle force of birds in flight, 35

X

X-ray diffraction, 230–234

Computer Science and Applied Mathematics

A SERIES OF MONOGRAPHS AND TEXTBOOKS

Editor
Werner Rheinboldt
University of Maryland

HANS P. KÜNZI, H. G. TZSCHACH, and C. A. ZEHNDER. Numerical Methods of Mathematical Optimization: With ALGOL and FORTRAN Programs, Corrected and Augmented Edition

AZRIEL ROSENFELD. Picture Processing by Computer

JAMES ORTEGA AND WERNER RHEINBOLDT. Iterative Solution of Nonlinear Equations in Several Variables

AZARIA PAZ. Introduction to Probabilistic Automata

DAVID YOUNG. Iterative Solution of Large Linear Systems

ANN YASUHARA. Recursive Function Theory and Logic

JAMES M. ORTEGA. Numerical Analysis: A Second Course

G. W. STEWART. Introduction to Matrix Computations

CHIN-LIANG CHANG AND RICHARD CHAR-TUNG LEE. Symbolic Logic and Mechanical Theorem Proving

C. C. GOTLIEB AND A. BORODIN. Social Issues in Computing

ERWIN ENGELER. Introduction to the Theory of Computation

F. W. J. OLVER. Asymptotics and Special Functions

DIONYSIOS C. TSICHRITZIS AND PHILIP A. BERNSTEIN. Operating Systems

ROBERT R. KORFHAGE. Discrete Computational Structures

PHILIP J. DAVIS AND PHILIP RABINOWITZ. Methods of Numerical Integration

A. T. BERZTISS. Data Structures: Theory and Practice, Second Edition

N. CHRISTOPHIDES. Graph Theory: An Algorithmic Approach

ALBERT NIJENHUIS AND HERBERT S. WILF. Combinatorial Algorithms

AZRIEL ROSENFELD AND AVINASH C. KAK. Digital Picture Processing

SAKTI P. GHOSH. Data Base Organization for Data Management

DIONYSIOS C. TSICHRITZIS AND FREDERICK H. LOCHOVSKY. Data Base Management Systems

JAMES L. PETERSON. Computer Organization and Assembly Language Programming

WILLIAM F. AMES. Numerical Methods for Partial Differential Equations, Second Edition

ARNOLD O. ALLEN. Probability, Statistics, and Queueing Theory: With Computer Science Applications

ELLIOTT I. ORGANICK, ALEXANDRA I. FORSYTHE, AND ROBERT P. PLUMMER. Programming Language Structures

ALBERT NIJENHUIS AND HERBERT S. WILF. Combinatorial Algorithms. Second edition.

JAMES S. VANDERGRAFT. Introduction to Numerical Computations

AZRIEL ROSENFELD. Picture Languages, Formal Models for Picture Recognition

ISAAC FRIED. Numerical Solution of Differential Equations

ABRAHAM BERMAN AND ROBERT J. PLEMMONS. Nonnegative Matrices in the Mathematical Sciences

BERNARD KOLMAN AND ROBERT E. BECK. Elementary Linear Programming with Applications

CLIVE L. DYM AND ELIZABETH S. IVEY. Principles of Mathematical Modeling

ERNEST L. HALL. Computer Image Processing and Recognition

ALLEN B. TUCKER, JR., Text Processing: Algorithms, Languages, and Applications

In preparation

MARTIN CHARLES GOLUMBIC. Algorithmic Graph Theory and Perfect Graphs

A
B
C
D 9
E 0
F 1
G 2
H 3
I 4
J 5